NONRESIDENT TRAINING COURSE

SEPTEMBER 1998

Navy Electricity and Electronics Training Series

Module 2—Introduction to Alternating Current and Transformers

NAVEDTRA 14174

DISTRIBUTION STATEMENT A: Approved for public release; distribution is unlimited.

Although the words "he," "him," and "his" are used sparingly in this course to enhance communication, they are not intended to be gender driven or to affront or discriminate against anyone.

DISTRIBUTION STATEMENT A: Approved for public release; distribution is unlimited.

PREFACE

By enrolling in this self-study course, you have demonstrated a desire to improve yourself and the Navy. Remember, however, this self-study course is only one part of the total Navy training program. Practical experience, schools, selected reading, and your desire to succeed are also necessary to successfully round out a fully meaningful training program.

COURSE OVERVIEW: To introduce the student to the subject of Alternating Current and Transformers who needs such a background in accomplishing daily work and/or in preparing for further study.

THE COURSE: This self-study course is organized into subject matter areas, each containing learning objectives to help you determine what you should learn along with text and illustrations to help you understand the information. The subject matter reflects day-to-day requirements and experiences of personnel in the rating or skill area. It also reflects guidance provided by Enlisted Community Managers (ECMs) and other senior personnel, technical references, instructions, etc., and either the occupational or naval standards, which are listed in the *Manual of Navy Enlisted Manpower Personnel Classifications and Occupational Standards*, NAVPERS 18068.

THE QUESTIONS: The questions that appear in this course are designed to help you understand the material in the text.

VALUE: In completing this course, you will improve your military and professional knowledge. Importantly, it can also help you study for the Navy-wide advancement in rate examination. If you are studying and discover a reference in the text to another publication for further information, look it up.

1998 Edition Prepared by
DSC Ray A. Jackson

Published by
NAVAL EDUCATION AND TRAINING
PROFESSIONAL DEVELOPMENT
AND TECHNOLOGY CENTER

NAVSUP Logistics Tracking Number
0504-LP-026-8270

Sailor's Creed

"I am a United States Sailor.

I will support and defend the Constitution of the United States of America and I will obey the orders of those appointed over me.

I represent the fighting spirit of the Navy and those who have gone before me to defend freedom and democracy around the world.

I proudly serve my country's Navy combat team with honor, courage and commitment.

I am committed to excellence and the fair treatment of all."

TABLE OF CONTENTS

CHAPTER   **PAGE**

1. Concepts of Alternating Current 1-1
2. Inductance 2-1
3. Capacitance 3-1
4. Inductive and Capacitive Reactance.. 4-1
5. Transformers 5-1

APPENDIX

    I. Glossary... AI-1

   II. Greek Alphabet.. AII-1

  III. Square and Square Roots.. AIII-1

  IV. Useful AC Formulas.. AIV-1

   V. Trigonometric Functions AV-1

 VI. Trigonometric Tables AVI-1

INDEX INDEX-1

NAVY ELECTRICITY AND ELECTRONICS TRAINING SERIES

The Navy Electricity and Electronics Training Series (NEETS) was developed for use by personnel in many electrical- and electronic-related Navy ratings. Written by, and with the advice of, senior technicians in these ratings, this series provides beginners with fundamental electrical and electronic concepts through self-study. The presentation of this series is not oriented to any specific rating structure, but is divided into modules containing related information organized into traditional paths of instruction.

The series is designed to give small amounts of information that can be easily digested before advancing further into the more complex material. For a student just becoming acquainted with electricity or electronics, it is highly recommended that the modules be studied in their suggested sequence. While there is a listing of NEETS by module title, the following brief descriptions give a quick overview of how the individual modules flow together.

Module 1, *Introduction to Matter, Energy, and Direct Current*, introduces the course with a short history of electricity and electronics and proceeds into the characteristics of matter, energy, and direct current (dc). It also describes some of the general safety precautions and first-aid procedures that should be common knowledge for a person working in the field of electricity. Related safety hints are located throughout the rest of the series, as well.

Module 2, *Introduction to Alternating Current and Transformers,* is an introduction to alternating current (ac) and transformers, including basic ac theory and fundamentals of electromagnetism, inductance, capacitance, impedance, and transformers.

Module 3, *Introduction to Circuit Protection, Control, and Measurement,* encompasses circuit breakers, fuses, and current limiters used in circuit protection, as well as the theory and use of meters as electrical measuring devices.

Module 4, *Introduction to Electrical Conductors, Wiring Techniques, and Schematic Reading,* presents conductor usage, insulation used as wire covering, splicing, termination of wiring, soldering, and reading electrical wiring diagrams.

Module 5, *Introduction to Generators and Motors,* is an introduction to generators and motors, and covers the uses of ac and dc generators and motors in the conversion of electrical and mechanical energies.

Module 6, *Introduction to Electronic Emission, Tubes, and Power Supplies,* ties the first five modules together in an introduction to vacuum tubes and vacuum-tube power supplies.

Module 7, *Introduction to Solid-State Devices and Power Supplies,* is similar to module 6, but it is in reference to solid-state devices.

Module 8, *Introduction to Amplifiers,* covers amplifiers.

Module 9, *Introduction to Wave-Generation and Wave-Shaping Circuits,* discusses wave generation and wave-shaping circuits.

Module 10, *Introduction to Wave Propagation, Transmission Lines, and Antennas,* presents the characteristics of wave propagation, transmission lines, and antennas.

Module 11, *Microwave Principles,* explains microwave oscillators, amplifiers, and waveguides.

Module 12, *Modulation Principles,* discusses the principles of modulation.

Module 13, *Introduction to Number Systems and Logic Circuits,* presents the fundamental concepts of number systems, Boolean algebra, and logic circuits, all of which pertain to digital computers.

Module 14, *Introduction to Microelectronics,* covers microelectronics technology and miniature and microminiature circuit repair.

Module 15, *Principles of Synchros, Servos, and Gyros,* provides the basic principles, operations, functions, and applications of synchro, servo, and gyro mechanisms.

Module 16, *Introduction to Test Equipment,* is an introduction to some of the more commonly used test equipments and their applications.

Module 17, *Radio-Frequency Communications Principles,* presents the fundamentals of a radio-frequency communications system.

Module 18, *Radar Principles,* covers the fundamentals of a radar system.

Module 19, *The Technician's Handbook,* is a handy reference of commonly used general information, such as electrical and electronic formulas, color coding, and naval supply system data.

Module 20, *Master Glossary,* is the glossary of terms for the series.

Module 21, *Test Methods and Practices,* describes basic test methods and practices.

Module 22, *Introduction to Digital Computers,* is an introduction to digital computers.

Module 23, *Magnetic Recording,* is an introduction to the use and maintenance of magnetic recorders and the concepts of recording on magnetic tape and disks.

Module 24, *Introduction to Fiber Optics,* is an introduction to fiber optics.

Embedded questions are inserted throughout each module, except for modules 19 and 20, which are reference books. If you have any difficulty in answering any of the questions, restudy the applicable section.

Although an attempt has been made to use simple language, various technical words and phrases have necessarily been included. Specific terms are defined in Module 20, *Master Glossary*.

Considerable emphasis has been placed on illustrations to provide a maximum amount of information. In some instances, a knowledge of basic algebra may be required.

Assignments are provided for each module, with the exceptions of Module 19, *The Technician's Handbook*; and Module 20, *Master Glossary*. Course descriptions and ordering information are in NAVEDTRA 12061, *Catalog of Nonresident Training Courses*.

Throughout the text of this course and while using technical manuals associated with the equipment you will be working on, you will find the below notations at the end of some paragraphs. The notations are used to emphasize that safety hazards exist and care must be taken or observed.

WARNING

AN OPERATING PROCEDURE, PRACTICE, OR CONDITION, ETC., WHICH MAY RESULT IN INJURY OR DEATH IF NOT CAREFULLY OBSERVED OR FOLLOWED.

CAUTION

AN OPERATING PROCEDURE, PRACTICE, OR CONDITION, ETC., WHICH MAY RESULT IN DAMAGE TO EQUIPMENT IF NOT CAREFULLY OBSERVED OR FOLLOWED.

NOTE

An operating procedure, practice, or condition, etc., which is essential to emphasize.

INSTRUCTIONS FOR TAKING THE COURSE

ASSIGNMENTS

The text pages that you are to study are listed at the beginning of each assignment. Study these pages carefully before attempting to answer the questions. Pay close attention to tables and illustrations and read the learning objectives. The learning objectives state what you should be able to do after studying the material. Answering the questions correctly helps you accomplish the objectives.

SELECTING YOUR ANSWERS

Read each question carefully, then select the BEST answer. You may refer freely to the text. The answers must be the result of your own work and decisions. You are prohibited from referring to or copying the answers of others and from giving answers to anyone else taking the course.

SUBMITTING YOUR ASSIGNMENTS

To have your assignments graded, you must be enrolled in the course with the Nonresident Training Course Administration Branch at the Naval Education and Training Professional Development and Technology Center (NETPDTC). Following enrollment, there are two ways of having your assignments graded: (1) use the Internet to submit your assignments as you complete them, or (2) send all the assignments at one time by mail to NETPDTC.

Grading on the Internet: Advantages to Internet grading are:

- you may submit your answers as soon as you complete an assignment, and
- you get your results faster; usually by the next working day (approximately 24 hours).

In addition to receiving grade results for each assignment, you will receive course completion confirmation once you have completed all the assignments. To submit your assignment answers via the Internet, go to:

http://courses.cnet.navy.mil

Grading by Mail: When you submit answer sheets by mail, send all of your assignments at one time. Do NOT submit individual answer sheets for grading. Mail all of your assignments in an envelope, which you either provide yourself or obtain from your nearest Educational Services Officer (ESO). Submit answer sheets to:

COMMANDING OFFICER
NETPDTC N331
6490 SAUFLEY FIELD ROAD
PENSACOLA FL 32559-5000

Answer Sheets: All courses include one "scannable" answer sheet for each assignment. These answer sheets are preprinted with your SSN, name, assignment number, and course number. Explanations for completing the answer sheets are on the answer sheet.

Do not use answer sheet reproductions: Use only the original answer sheets that we provide—reproductions will not work with our scanning equipment and cannot be processed.

Follow the instructions for marking your answers on the answer sheet. Be sure that blocks 1, 2, and 3 are filled in correctly. This information is necessary for your course to be properly processed and for you to receive credit for your work.

COMPLETION TIME

Courses must be completed within 12 months from the date of enrollment. This includes time required to resubmit failed assignments.

PASS/FAIL ASSIGNMENT PROCEDURES

If your overall course score is 3.2 or higher, you will pass the course and will not be required to resubmit assignments. Once your assignments have been graded you will receive course completion confirmation.

If you receive less than a 3.2 on any assignment and your overall course score is below 3.2, you will be given the opportunity to resubmit failed assignments. **You may resubmit failed assignments only once.** Internet students will receive notification when they have failed an assignment--they may then resubmit failed assignments on the web site. Internet students may view and print results for failed assignments from the web site. Students who submit by mail will receive a failing result letter and a new answer sheet for resubmission of each failed assignment.

COMPLETION CONFIRMATION

After successfully completing this course, you will receive a letter of completion.

ERRATA

Errata are used to correct minor errors or delete obsolete information in a course. Errata may also be used to provide instructions to the student. If a course has an errata, it will be included as the first page(s) after the front cover. Errata for all courses can be accessed and viewed/downloaded at:

http://www.advancement.cnet.navy.mil

STUDENT FEEDBACK QUESTIONS

We value your suggestions, questions, and criticisms on our courses. If you would like to communicate with us regarding this course, we encourage you, if possible, to use e-mail. If you write or fax, please use a copy of the Student Comment form that follows this page.

For subject matter questions:

E-mail: n315.products@cnet.navy.mil
Phone: Comm: (850) 452-1001, ext. 1728
DSN: 922-1001, ext. 1728
FAX: (850) 452-1370
(Do not fax answer sheets.)
Address: COMMANDING OFFICER
NETPDTC N315
6490 SAUFLEY FIELD ROAD
PENSACOLA FL 32509-5237

For enrollment, shipping, grading, or completion letter questions

E-mail: fleetservices@cnet.navy.mil
Phone: Toll Free: 877-264-8583
Comm: (850) 452-1511/1181/1859
DSN: 922-1511/1181/1859
FAX: (850) 452-1370
(Do not fax answer sheets.)
Address: COMMANDING OFFICER
NETPDTC N331
6490 SAUFLEY FIELD ROAD
PENSACOLA FL 32559-5000

NAVAL RESERVE RETIREMENT CREDIT

If you are a member of the Naval Reserve, you will receive retirement points if you are authorized to receive them under current directives governing retirement of Naval Reserve personnel. For Naval Reserve retirement, this course is evaluated at 10 points. (Refer to *Administrative Procedures for Naval Reservists on Inactive Duty,* BUPERSINST 1001.39, for more information about retirement points.)

Student Comments

Course Title: *NEETS Module 2*
Introduction to Alternating Current and Transformers

NAVEDTRA: 14174   **Date:** _____

We need some information about you:

Rate/Rank and Name: _____ SSN: _____ Command/Unit _____

Street Address: _____ City: _____ State/FPO: _____ Zip _____

Your comments, suggestions, etc.:

Privacy Act Statement: Under authority of Title 5, USC 301, information regarding your military status is requested in processing your comments and in preparing a reply. This information will not be divulged without written authorization to anyone other than those within DOD for official use in determining performance.

NETPDTC 1550/41 (Rev 4-00)

CHAPTER 1
CONCEPTS OF ALTERNATING CURRENT

LEARNING OBJECTIVES

Upon completion of this chapter you will be able to:

1. State the differences between ac and dc voltage and current.

2. State the advantages of ac power transmission over dc power transmission.

3. State the "left-hand rule" for a conductor.

4. State the relationship between current and magnetism.

5. State the methods by which ac power can be generated.

6. State the relationship between frequency, period, time, and wavelength.

7. Compute peak-to-peak, instantaneous, effective, and average values of voltage and current.

8. Compute the phase difference between sine waves.

CONCEPTS OF ALTERNATING CURRENT

All of your study thus far has been with direct current (dc), that is, current which does not change direction. However, as you saw in module 1 and will see later in this module, a coil rotating in a magnetic field actually generates a current which regularly changes direction. This current is called ALTERNATING CURRENT or ac.

AC AND DC

Alternating current is current which constantly changes in amplitude, and which reverses direction at regular intervals. You learned previously that direct current flows only in one direction, and that the amplitude of current is determined by the number of electrons flowing past a point in a circuit in one second. If, for example, a coulomb of electrons moves past a point in a wire in one second and all of the electrons are moving in the same direction, the amplitude of direct current in the wire is one ampere. Similarly, if half a coulomb of electrons moves in one direction past a point in the wire in half a second, then reverses direction and moves past the same point in the opposite direction during the next half-second, a total of one coulomb of electrons passes the point in one second. The amplitude of the alternating current is one ampere. The preceding comparison of dc and ac as illustrated. Notice that one white arrow plus one striped arrow comprise one coulomb.

COMPARING DC & AC CURRENT FLOW IN A WIRE

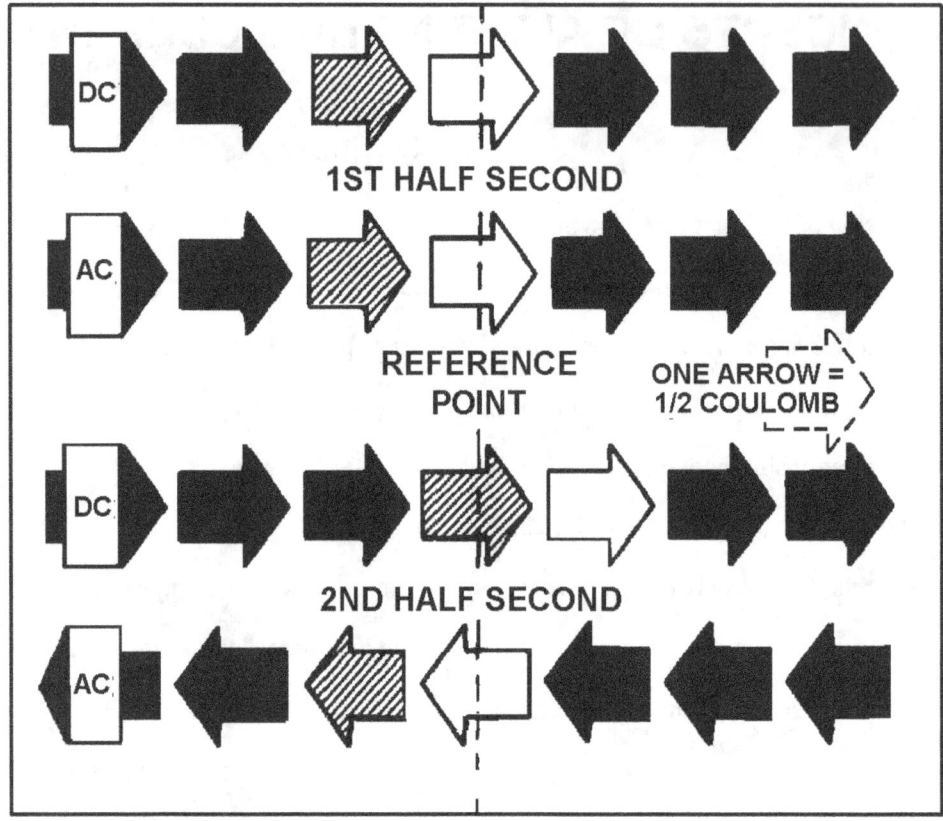

Q1. Define direct current.

Q2. Define alternating current.

DISADVANTAGES OF DC COMPARED TO AC

When commercial use of electricity became wide-spread in the United States, certain disadvantages in using direct current in the home became apparent. If a commercial direct-current system is used, the voltage must be generated at the level (amplitude or value) required by the load. To properly light a 240-volt lamp, for example, the dc generator must deliver 240 volts. If a 120-volt lamp is to be supplied power from the 240-volt generator, a resistor or another 120-volt lamp must be placed in series with the 120-volt lamp to drop the extra 120 volts. When the resistor is used to reduce the voltage, an amount of power equal to that consumed by the lamp is wasted.

Another disadvantage of the direct-current system becomes evident when the direct current (I) from the generating station must be transmitted a long distance over wires to the consumer. When this happens, a large amount of power is lost due to the resistance (R) of the wire. The power loss is equal to I^2R. However, this loss can be greatly reduced if the power is transmitted over the lines at a very high voltage level and a low current level. This is not a practical solution to the power loss in the dc system since the load would then have to be operated at a dangerously high voltage. Because of the disadvantages related to transmitting and using direct current, practically all modern commercial electric power companies generate and distribute alternating current (ac).

Unlike direct voltages, alternating voltages can be stepped up or down in amplitude by a device called a TRANSFORMER. (The transformer will be explained later in this module.) Use of the transformer permits efficient transmission of electrical power over long-distance lines. At the electrical power station, the transformer output power is at high voltage and low current levels. At the consumer end of the transmission lines, the voltage is stepped down by a transformer to the value required by the load. Due to its inherent advantages and versatility, alternating current has replaced direct current in all but a few commercial power distribution systems.

Q3. *What is a disadvantage of a direct-current system with respect to supply voltage?*

Q4. *What disadvantage of a direct current is due to the resistance of the transmission wires?*

Q5. *What kind of electrical current is used in most modern power distribution systems?*

VOLTAGE WAVEFORMS

You now know that there are two types of current and voltage, that is, direct current and voltage and alternating current and voltage. If a graph is constructed showing the amplitude of a dc voltage across the terminals of a battery with respect to time, it will appear in figure 1-1 view A. The dc voltage is shown to have a constant amplitude. Some voltages go through periodic changes in amplitude like those shown in figure 1-1 view B. The pattern which results when these changes in amplitude with respect to time are plotted on graph paper is known as a WAVEFORM. Figure 1-1 view B shows some of the common electrical waveforms. Of those illustrated, the sine wave will be dealt with most often.

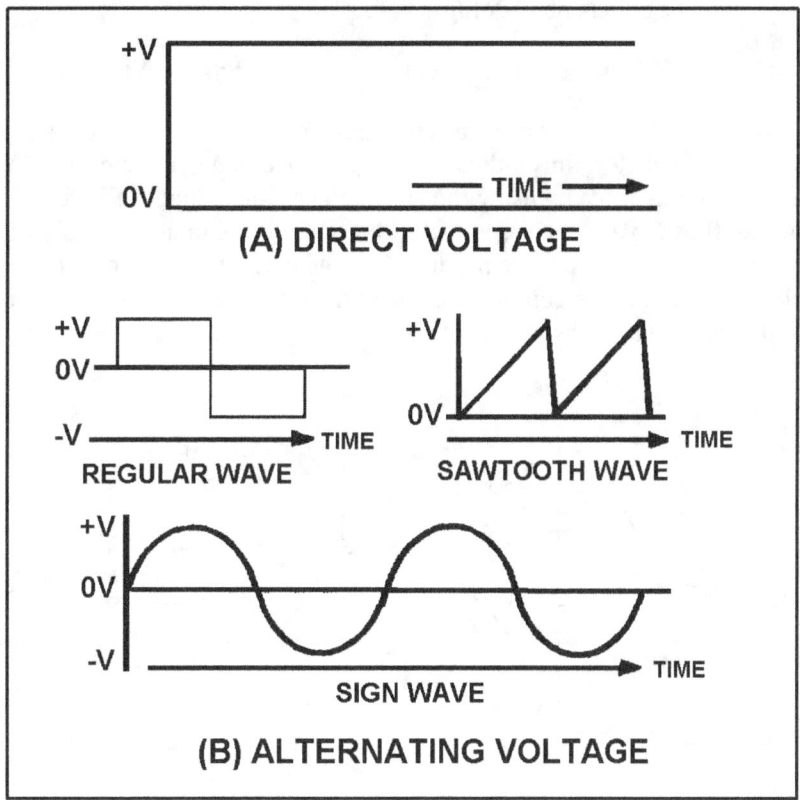

Figure 1-1.—Voltage waveforms: (A) Direct voltage; (B) Alternating voltage.

ELECTROMAGNETISM

The sine wave illustrated in figure 1-1 view B is a plot of a current which changes amplitude and direction. Although there are several ways of producing this current, the method based on the principles of electromagnetic induction is by far the easiest and most common method in use.

The fundamental theories concerning simple magnets and magnetism were discussed in Module 1, but how magnetism can be used to produce electricity was only briefly mentioned. This module will give you a more in-depth study of magnetism. The main points that will be explained are how magnetism is affected by an electric current and, conversely, how electricity is affected by magnetism. This general subject area is most often referred to as ELECTROMAGNETISM. To properly understand electricity you must first become familiar with the relationships between magnetism and electricity. For example, you must know that:

- An electric current always produces some form of magnetism.

- The most commonly used means for producing or using electricity involves magnetism.

- The peculiar behavior of electricity under certain conditions is caused by magnetic influences.

MAGNETIC FIELDS

In 1819 Hans Christian Oersted, a Danish physicist, found that a definite relationship exists between magnetism and electricity. He discovered that an electric current is always accompanied by certain magnetic effects and that these effects obey definite laws.

MAGNETIC FIELD AROUND A CURRENT-CARRYING CONDUCTOR

If a compass is placed in the vicinity of a current-carrying conductor, the compass needle will align itself at right angles to the conductor, thus indicating the presence of a magnetic force. You can demonstrate the presence of this force by using the arrangement illustrated in figure 1-2. In both (A) and (B) of the figure, current flows in a vertical conductor through a horizontal piece of cardboard. You can determine the direction of the magnetic force produced by the current by placing a compass at various points on the cardboard and noting the compass needle deflection. The direction of the magnetic force is assumed to be the direction in which the north pole of the compass points.

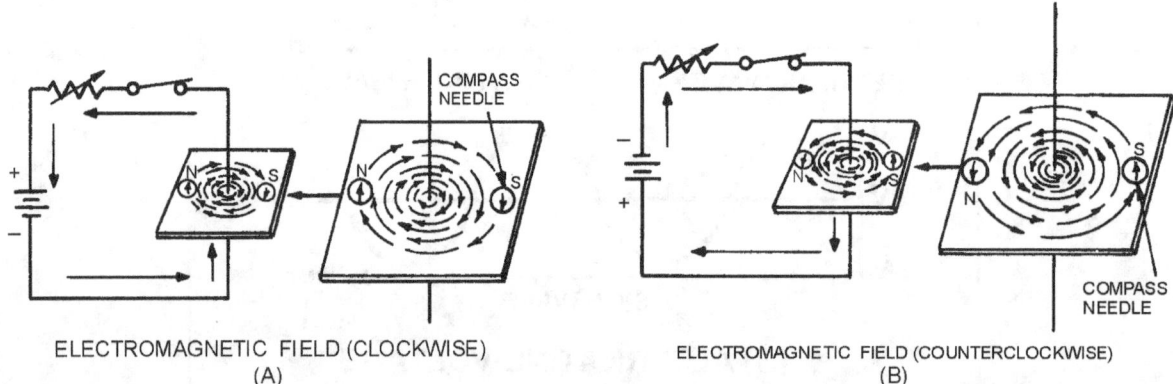

Figure 1-2.—Magnetic field around a current-carrying conductor.

In figure 1-2 (A), the needle deflections show that a magnetic field exists in circular form around the conductor. When the current flows upward (see figure 1-2(A)), the direction of the field is clockwise, as viewed from the top. However, if you reverse the polarity of the battery so that the current flows downward (see figure 1-2(B)), the direction of the field is counterclockwise.

The relation between the direction of the magnetic lines of force around a conductor and the direction of electron current flow in the conductor may be determined by means of the LEFT-HAND RULE FOR A CONDUCTOR: if you grasp the conductor in your left hand with the thumb extended in the direction of the electron flow (current) (– to +), your fingers will point in the direction of the magnetic lines of force. Now apply this rule to figure 1-2. Note that your fingers point in the direction that the north pole of the compass points when it is placed in the magnetic field surrounding the wire.

An arrow is generally used in electrical diagrams to denote the direction of current in a length of wire (see figure 1-3(A)). Where a cross section of a wire is shown, an end view of the arrow is used. A cross-sectional view of a conductor that is carrying current toward the observer is illustrated in figure 1-3(B). Notice that the direction of current is indicated by a dot, representing the head of the arrow. A conductor that is carrying current away from the observer is illustrated in figure 1-3(C). Note that the direction of current is indicated by a cross, representing the tail of the arrow. Also note that the magnetic field around a current-carrying conductor is perpendicular to the conductor, and that the magnetic lines of force are equal along all parts of the conductor.

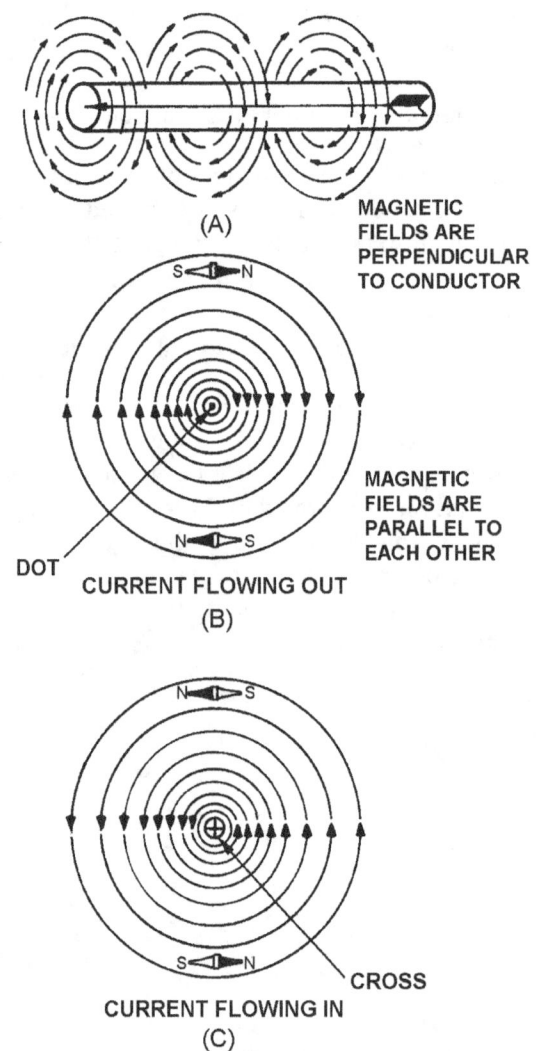

Figure 1-3.—Magnetic field around a current-carrying conductor, detailed view.

When two adjacent parallel conductors are carrying current in the same direction, the magnetic lines of force combine and increase the strength of the field around the conductors, as shown in figure 1-4(A). Two parallel conductors carrying currents in opposite directions are shown in figure 1-4(B). Note that the field around one conductor is opposite in direction to the field around the other conductor. The resulting lines of force oppose each other in the space between the wires, thus deforming the field around each conductor. This means that if two parallel and adjacent conductors are carrying currents in the same direction, the fields about the two conductors aid each other. Conversely, if the two conductors are carrying currents in opposite directions, the fields about the conductors repel each other.

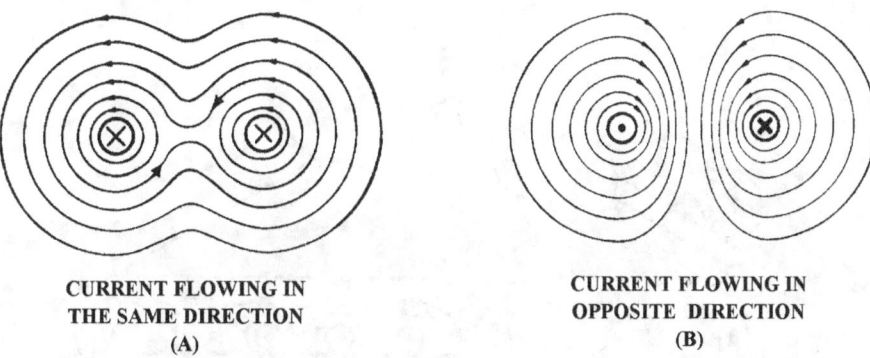

Figure 1-4.—Magnetic field around two parallel conductors.

Q6. When placed in the vicinity of a current-carrying conductor, the needle of a compass becomes aligned at what angle to the conductor?

Q7. What is the direction of the magnetic field around a vertical conductor when (a) the current flows upward and (b) the current flows downward.

Q8. The "left-hand rule" for a conductor is used for what purpose

Q9. In what direction will the compass needle point when the compass is placed in the magnetic field surrounding a wire?

Q10. When two adjacent parallel wires carry current in the same direction, the magnetic field about one wire has what effect on the magnetic field about the other conductor?

Q11. When two adjacent parallel conductors carry current in opposite directions, the magnetic field about one conductor has what effect on the magnetic field about the other conductor?

MAGNETIC FIELD OF A COIL

Figure 1-3(A) illustrates that the magnetic field around a current-carrying wire exists at all points along the wire. Figure 1-5 illustrates that when a straight wire is wound around a core, it forms a coil and that the magnetic field about the core assumes a different shape. Figure 1-5(A) is actually a partial cutaway view showing the construction of a simple coil. Figure 1-5(B) shows a cross-sectional view of the same coil. Notice that the two ends of the coil are identified as X and Y.

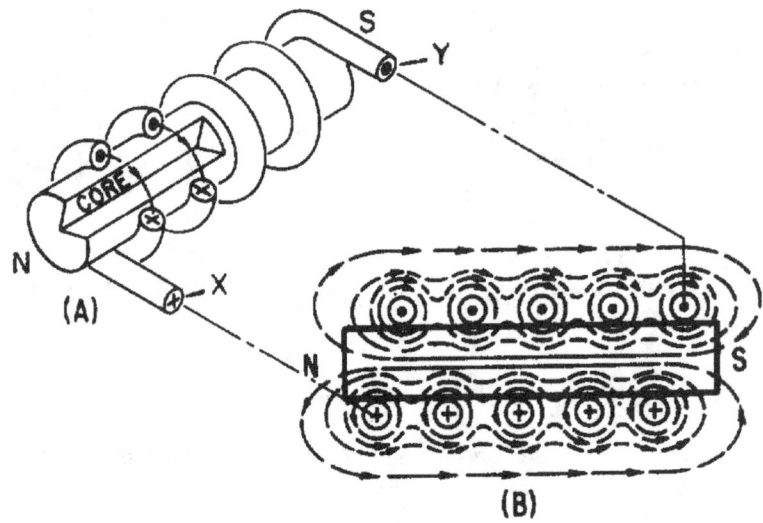

Figure 1-5.—Magnetic field produced by a current-carrying coil.

When current is passed through the coil, the magnetic field about each turn of wire links with the fields of the adjacent turns. (See figure 1-4(A)). The combined influence of all the turns produces a two-pole field similar to that of a simple bar magnet. One end of the coil is a north pole and the other end is a south pole.

Polarity of an Electromagnetic Coil

Figure 1-2 shows that the direction of the magnetic field around a straight wire depends on the direction of current in that wire. Thus, a reversal of current in a wire causes a reversal in the direction of the magnetic field that is produced. It follows that a reversal of the current in a coil also causes a reversal of the two-pole magnetic field about the coil.

When the direction of the current in a coil is known, you can determine the magnetic polarity of the coil by using the LEFT-HAND RULE FOR COILS. This rule, illustrated in figure 1-6, is stated as follows:

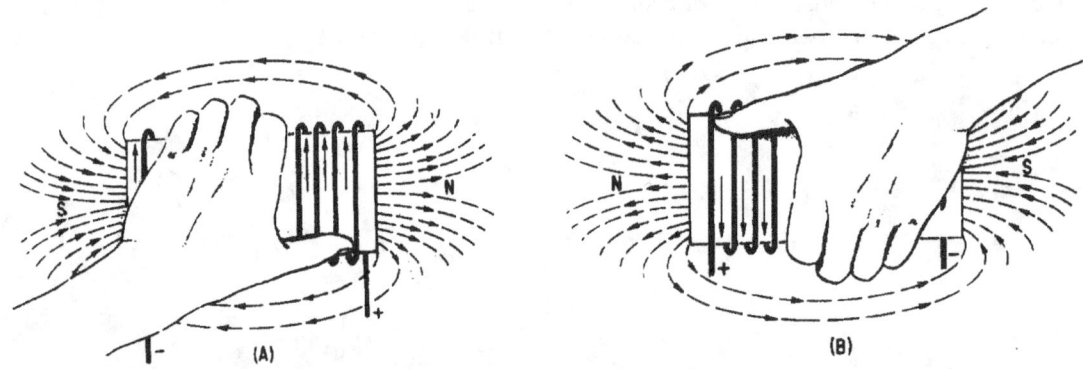

Figure 1-6.—Left-hand rule for coils.

Grasp the coil in your left hand, with your fingers "wrapped around" in the direction of the electron current flow. Your thumb will then point toward the north pole of the coil.

Strength of an Electromagnetic Field

The strength or intensity of a coil's magnetic field depends on a number of factors. The main ones are listed below and will be discussed again later.

- The number of turns of wire in the coil.
- The amount of current flowing in the coil.
- The ratio of the coil length to the coil width.
- The type of material in the core.

Losses in an Electromagnetic Field

When current flows in a conductor, the atoms in the conductor all line up in a definite direction, producing a magnetic field. When the direction of the current changes, the direction of the atoms' alignment also changes, causing the magnetic field to change direction. To reverse all the atoms requires that power be expended, and this power is lost. This loss of power (in the form of heat) is called HYSTERESIS LOSS. Hysteresis loss is common to all ac equipment; however, it causes few problems except in motors, generators, and transformers. When these devices are discussed later in this module, hysteresis loss will be covered in more detail.

Q12. What is the shape of the magnetic field that exists around (a) a straight conductor and (b) a coil?

Q13. What happens to the two-pole field of a coil when the current through the coil is reversed?

Q14. What rule is used to determine the polarity of a coil when the direction of the electron current flow in the coil is known?

Q15. State the rule whose purpose is described in Q14.

BASIC AC GENERATION

From the previous discussion you learned that a current-carrying conductor produces a magnetic field around itself. In module 1, under producing a voltage (emf) using magnetism, you learned how a changing magnetic field produces an emf in a conductor. That is, if a conductor is placed in a magnetic field, and either the field or the conductor moves, an emf is induced in the conductor. This effect is called electromagnetic induction.

CYCLE

Figures 1-7 and 1-8 show a suspended loop of wire (conductor) being rotated (moved) in a clockwise direction through the magnetic field between the poles of a permanent magnet. For ease of explanation, the loop has been divided into a dark half and light half. Notice in (A) of the figure that the dark half is moving along (parallel to) the lines of force. Consequently, it is cutting NO lines of force. The same is true of the light half, which is moving in the opposite direction. Since the conductors are cutting no lines of force, no emf is induced. As the loop rotates toward the position shown in (B), it cuts more and more lines of force per second (inducing an ever-increasing voltage) because it is cutting more directly across the field (lines of force). At (B), the conductor is shown completing one-quarter of a complete revolution, or 90°, of a complete circle. Because the conductor is now cutting directly across the field, the voltage induced in the conductor is maximum. When the value of induced voltage at various points during the rotation from (A) to (B) is plotted on a graph (and the points connected), a curve appears as shown below.

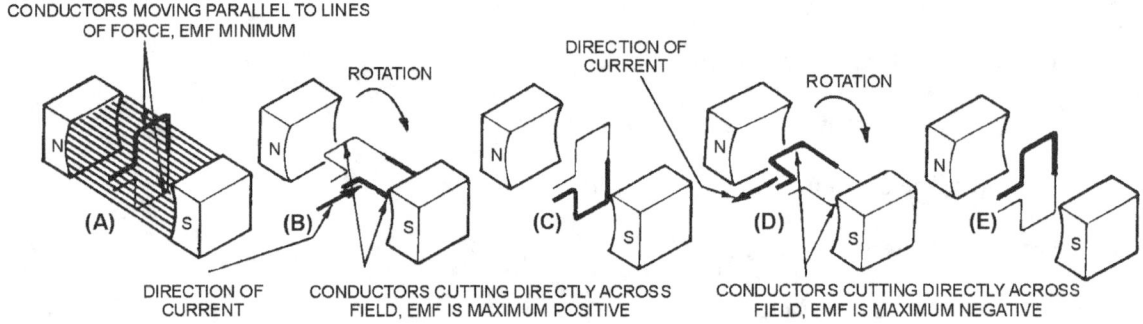

Figure 1-7.—Simple alternating-current generator.

As the loop continues to be rotated toward the position shown below in (C), it cuts fewer and fewer lines of force. The induced voltage decreases from its peak value. Eventually, the loop is once again moving in a plane parallel to the magnetic field, and no emf is induced in the conductor.

The loop has now been rotated through half a circle (one alternation or 180°). If the preceding quarter-cycle is plotted, it appears as shown below.

When the same procedure is applied to the second half of rotation (180° through 360°), the curve appears as shown below. Notice the only difference is in the polarity of the induced voltage. Where previously the polarity was positive, it is now negative.

The sine curve shows the value of induced voltage at each instant of time during rotation of the loop. Notice that this curve contains 360°, or two alternations. TWO ALTERNATIONS represent ONE complete CYCLE of rotation.

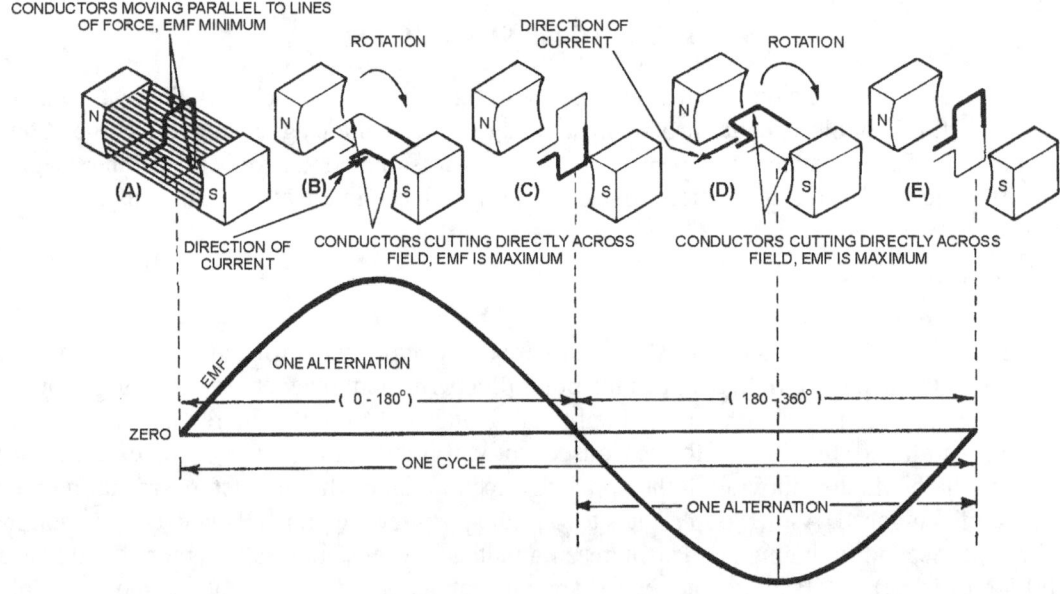

Figure 1-8.—Basic alternating-current generator.

1-10

Assuming a closed path is provided across the ends of the conductor loop, you can determine the direction of current in the loop by using the LEFT-HAND RULE FOR GENERATORS. Refer to figure 1-9. The left-hand rule is applied as follows: First, place your left hand on the illustration with the fingers as shown. Your THUMB will now point in the direction of rotation (relative movement of the wire to the magnetic field); your FOREFINGER will point in the direction of magnetic flux (north to south); and your MIDDLE FINGER (pointing out of the paper) will point in the direction of electron current flow.

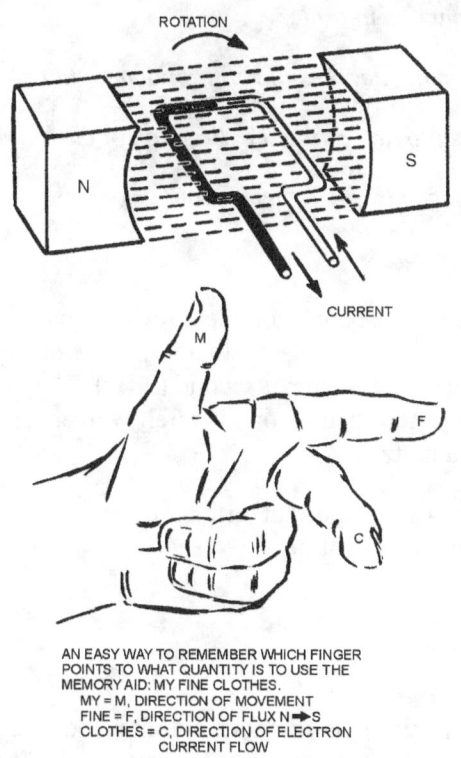

AN EASY WAY TO REMEMBER WHICH FINGER POINTS TO WHAT QUANTITY IS TO USE THE MEMORY AID: MY FINE CLOTHES.
MY = M, DIRECTION OF MOVEMENT
FINE = F, DIRECTION OF FLUX N→S
CLOTHES = C, DIRECTION OF ELECTRON CURRENT FLOW

Figure 1-9.—Left-hand rule for generators.

By applying the left-hand rule to the dark half of the loop in (B) in figure 1-8, you will find that the current flows in the direction indicated by the heavy arrow. Similarly, by using the left-hand rule on the light half of the loop, you will find that current therein flows in the opposite direction. The two induced voltages in the loop add together to form one total emf. It is this emf which causes the current in the loop.

When the loop rotates to the position shown in (D) of figure 1-8, the action reverses. The dark half is moving up instead of down, and the light half is moving down instead of up. By applying the left-hand rule once again, you will see that the total induced emf and its resulting current have reversed direction. The voltage builds up to maximum in this new direction, as shown by the sine curve in figure 1-8. The loop finally returns to its original position (E), at which point voltage is again zero. The sine curve represents one complete cycle of voltage generated by the rotating loop. All the illustrations used in this chapter show the wire loop moving in a clockwise direction. In actual practice, the loop can be moved clockwise or counterclockwise. Regardless of the direction of movement, the left-hand rule applies.

If the loop is rotated through 360° at a steady rate, and if the strength of the magnetic field is uniform, the voltage produced is a sine wave of voltage, as indicated in figure 1-9. Continuous rotation of the loop will produce a series of sine-wave voltage cycles or, in other words, an ac voltage.

As mentioned previously, the cycle consists of two complete alternations in a period of time. Recently the HERTZ (Hz) has been designated to indicate one cycle per second. If ONE CYCLE PER SECOND is ONE HERTZ, then 100 cycles per second are equal to 100 hertz, and so on. Throughout the NEETS, the term cycle is used when no specific time element is involved, and the term hertz (Hz) is used when the time element is measured in seconds.

Q16. When a conductor is rotated in a magnetic field, at what points in the cycle is emf (a) at maximum amplitude and (b) at minimum amplitude?

Q17. One cycle is equal to how many degrees of rotation of a conductor in a magnetic field?

Q18. State the left-hand rule used to determine the direction of current in a generator.

Q19. How is an ac voltage produced by an ac generator?

FREQUENCY

If the loop in the figure 1-8 (A) makes one complete revolution each second, the generator produces one complete cycle of ac during each second (1 Hz). Increasing the number of revolutions to two per second will produce two complete cycles of ac per second (2 Hz). The number of complete cycles of alternating current or voltage completed each second is referred to as the FREQUENCY. Frequency is always measured and expressed in hertz.

Alternating-current frequency is an important term to understand since most ac electrical equipments require a specific frequency for proper operation.

Q20. Define Frequency.

PERIOD

An individual cycle of any sine wave represents a definite amount of TIME. Notice that figure 1-10 shows 2 cycles of a sine wave which has a frequency of 2 hertz (Hz). Since 2 cycles occur each second, 1 cycle must require one-half second of time. The time required to complete one cycle of a waveform is called the PERIOD of the wave. In figure 1-10, the period is one-half second. The relationship between time (t) and frequency (f) is indicated by the formulas

$$t = \frac{1}{f} \text{ and } f = \frac{1}{t}$$

where t = period in seconds and
f = frequency in hertz

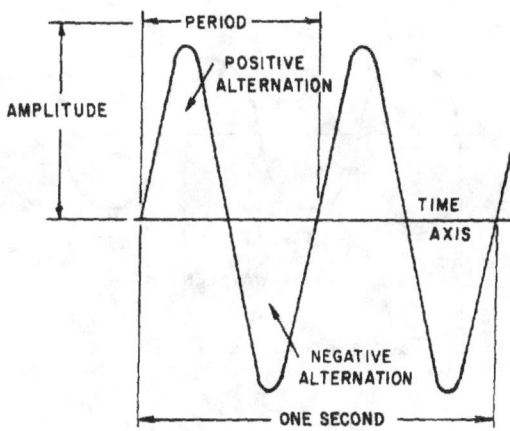

Figure 1-10.—Period of a sine wave.

Each cycle of the sine wave shown in figure 1-10 consists of two identically shaped variations in voltage. The variation which occurs during the time the voltage is positive is called the POSITIVE ALTERNATION. The variation which occurs during the time the voltage is negative is called the NEGATIVE ALTERNATION. In a sine wave, these two alternations are identical in size and shape, but opposite in polarity.

The distance from zero to the maximum value of each alternation is called the AMPLITUDE. The amplitude of the positive alternation and the amplitude of the negative alternation are the same.

WAVELENGTH

The time it takes for a sine wave to complete one cycle is defined as the period of the waveform. The distance traveled by the sine wave during this period is referred to as WAVELENGTH. Wavelength, indicated by the symbol λ (Greek lambda), is the distance along the waveform from one point to the same point on the next cycle. You can observe this relationship by examining figure 1-11. The point on the waveform that measurement of wavelength begins is not important as long as the distance is measured to the same point on the next cycle (see figure 1-12).

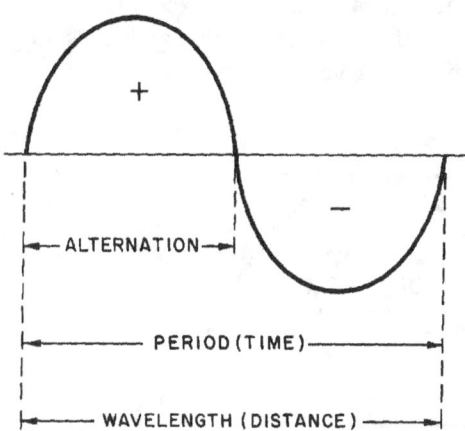

Figure 1-11.—Wavelength.

1-13

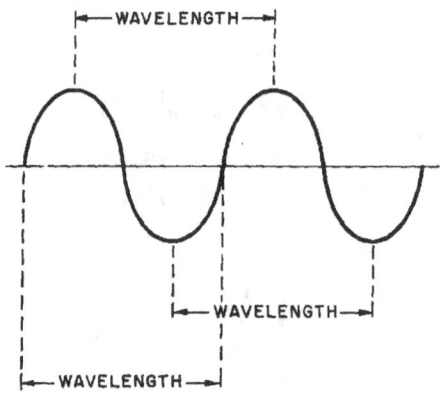

Figure 1-12.—Wavelength measurement.

Q21. What term is used to indicate the time of one complete cycle of a waveform?

Q22. What is a positive alternation?

Q23. What do the period and the wavelength of a sine wave measure, respectively?

ALTERNATING CURRENT VALUES

In discussing alternating current and voltage, you will often find it necessary to express the current and voltage in terms of MAXIMUM or PEAK values, PEAK-to-PEAK values, EFFECTIVE values, AVERAGE values, or INSTANTANEOUS values. Each of these values has a different meaning and is used to describe a different amount of current or voltage.

PEAK AND PEAK-TO-PEAK VALUES

Refer to figure 1-13. Notice it shows the positive alternation of a sine wave (a half-cycle of ac) and a dc waveform that occur simultaneously. Note that the dc starts and stops at the same moment as does the positive alternation, and that both waveforms rise to the same maximum value. However, the dc values are greater than the corresponding ac values at all points except the point at which the positive alternation passes through its maximum value. At this point the dc and ac values are equal. This point on the sine wave is referred to as the maximum or peak value.

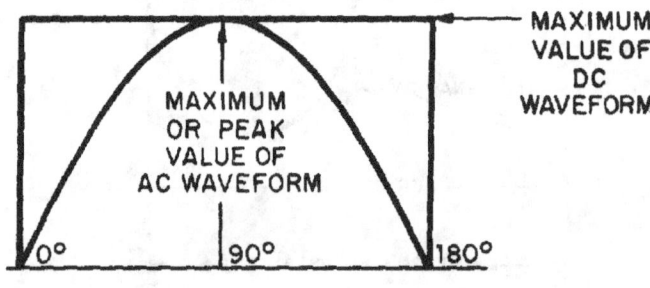

Figure 1-13.—Maximum or peak value.

1-14

During each complete cycle of ac there are always two maximum or peak values, one for the positive half-cycle and the other for the negative half-cycle. The difference between the peak positive value and the peak negative value is called the peak-to-peak value of the sine wave. This value is twice the maximum or peak value of the sine wave and is sometimes used for measurement of ac voltages. Note the difference between peak and peak-to-peak values in figure 1-14. Usually alternating voltage and current are expressed in EFFECTIVE VALUES (a term you will study later) rather than in peak-to-peak values.

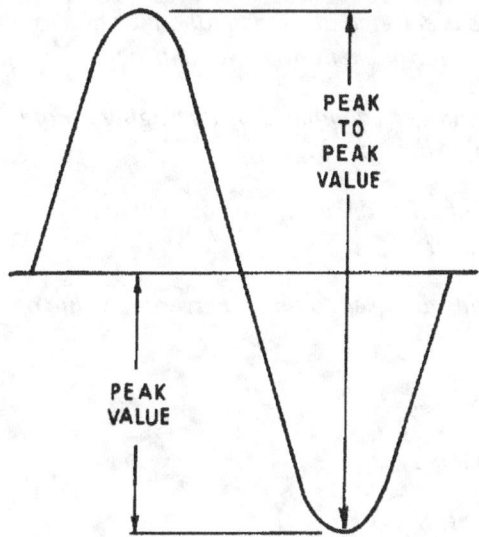

Figure 1-14.—Peak and peak-to-peak values.

Q24. What is meant by peak and peak-to-peak values of ac?

Q25. How many times is the maximum or peak value of emf or current reached during one cycle of ac?

INSTANTANEOUS VALUE

The INSTANTANEOUS value of an alternating voltage or current is the value of voltage or current at one particular instant. The value may be zero if the particular instant is the time in the cycle at which the polarity of the voltage is changing. It may also be the same as the peak value, if the selected instant is the time in the cycle at which the voltage or current stops increasing and starts decreasing. There are actually an infinite number of instantaneous values between zero and the peak value.

AVERAGE VALUE

The AVERAGE value of an alternating current or voltage is the average of ALL the INSTANTANEOUS values during ONE alternation. Since the voltage increases from zero to peak value and decreases back to zero during one alternation, the average value must be some value between those two limits. You could determine the average value by adding together a series of instantaneous values of the alternation (between 0° and 180°), and then dividing the sum by the number of instantaneous values used. The computation would show that one alternation of a sine wave has an average value equal to 0.636 times the peak value. The formula for average voltage is

$$E_{avg} = 0.636 \times E_{max}$$

where E_{avg} is the average voltage of one alternation, and E_{max} is the maximum or peak voltage. Similarly, the formula for average current is

$$I_{avg} = 0.636 \times I_{max}$$

where I_{avg} is the average current in one alternation, and I_{max} is the maximum or peak current.

Do not confuse the above definition of an average value with that of the average value of a complete cycle. Because the voltage is positive during one alternation and negative during the other alternation, the average value of the voltage values occurring during the complete cycle is <u>zero</u>.

Q26. *If any point on a sine wave is selected at random and the value of the current or voltage is measured at that one particular moment, what value is being measured?*

Q27. *What value of current or voltage is computed by averaging all of the instantaneous values during the negative alternation of a sine wave?*

Q28. *What is the average value of all of the instantaneous currents or voltages occurring during one complete cycle of a sine wave?*

Q29. *What mathematical formulas are used to find the average value of current and average value of voltage of a sine wave?*

Q30. *If E_{max} is 115 volts, what is E_{avg}?*

Q31. *If I_{avg} is 1.272 ampere, what is I_{max}?*

EFFECTIVE VALUE OF A SINE WAVE

E_{max}, E_{avg}, I_{max}, and I_{avg} are values used in ac measurements. Another value used is the EFFECTIVE value of ac This is the value of alternating voltage or current that will have the same effect on a resistance as a comparable value of direct voltage or current will have on the same resistance.

In an earlier discussion you were told that when current flows in a resistance, heat is produced. When direct current flows in a resistance, the amount of electrical power converted into heat equals I^2R watts. However, since an alternating current having a maximum value of 1 ampere does not maintain a constant value, the alternating current will not produce as much heat in the resistance as will a direct current of 1 ampere.

Figure 1-15 compares the heating effect of 1 ampere of dc to the heating effect of 1 ampere of ac.

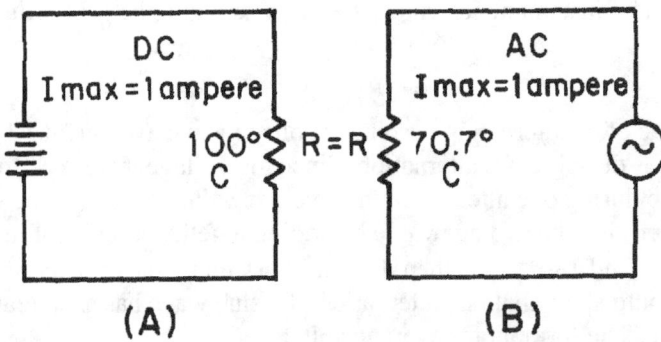

Figure 1-15.—Heating effect of ac and dc.

Examine views A and B of figure 1-15 and notice that the heat (70.7° C) produced by 1 ampere of alternating current (that is, an ac with a maximum value of 1 ampere) is only 70.7 percent of the heat (100° C) produced by 1 ampere of direct current. Mathematically,

$$\frac{\text{The heating effect of 1 maximum a.c. ampere}}{\text{The heating effect of 1 maximum d.c. ampere}} = \frac{70.7°\text{C}}{100°\text{C}} = 0.707$$

Therefore, for effective value of ac $(I_{eff}) = 0.707 \times I_{max}$.

The rate at which heat is produced in a resistance forms a convenient basis for establishing an effective value of alternating current, and is known as the "heating effect" method. An alternating current is said to have an effective value of one ampere when it produces heat in a given resistance at the same rate as does one ampere of direct current.

You can compute the effective value of a sine wave of current to a fair degree of accuracy by taking equally-spaced instantaneous values of current along the curve and extracting the square root of the average of the sum of the squared values.

For this reason, the effective value is often called the "root-mean-square" (rms) value. Thus,

$$I_{eff} = \sqrt{\text{Average of the sum of the squares of } I_{inst}}.$$

Stated another way, the effective or rms value (I_{eff}) of a sine wave of current is 0.707 times the maximum value of current (I_{max}). Thus, $I_{eff} = 0.707 \times I_{max}$. When I_{eff} is known, you can find I_{max} by using the formula $I_{max} = 1.414 \times I_{eff}$. You might wonder where the constant 1.414 comes from. To find out, examine figure 1-15 again and read the following explanation. Assume that the dc in figure 1-15(A) is maintained at 1 ampere and the resistor temperature at 100° C. Also assume that the ac in figure 1-15(B) is increased until the temperature of the resistor is 100° C. At this point it is found that a maximum ac value of 1.414 amperes is required in order to have the same heating effect as direct current. Therefore, in the ac circuit the maximum current required is 1.414 times the effective current. It is important for you to remember the above relationship and that the effective value (I_{eff}) of any sine wave of current is always 0.707 times the maximum value (I_{max}).

Since alternating current is caused by an alternating voltage, the ratio of the effective value of voltage to the maximum value of voltage is the same as the ratio of the effective value of current to the maximum value of current. Stated another way, the effective or rms value (E_{eff}) of a sine-wave of voltage is 0.707 times the maximum value of voltage (E_{max}),

Thus,

$$E_{eff} = \sqrt{\text{Average of the sum of the squares of } E_{inst}}$$

or,

$$E_{eff} = 0.707 \times E_{max}$$

and,

$$E_{max} = 1.414 \times E_{eff}$$

When an alternating current or voltage value is specified in a book or on a diagram, the value is an effective value unless there is a definite statement to the contrary. Remember that all meters, unless marked to the contrary, are calibrated to indicate effective values of current and voltage.

Problem: A circuit is known to have an alternating voltage of 120 volts and a peak or maximum current of 30 amperes. What are the peak voltage and effective current values?

Given:
$$E_s = 120 \text{ V}$$
$$E_{max} = 30 \text{ A}$$

Solution:
$$E_{max} = 1.414 \times E_{eff}$$
$$E_{max} = 1.414 \times 120 \text{ volts}$$
$$E_{max} = 169.68 \text{ volts}$$
$$I_{eff} = 0.707 \times I_{max}$$
$$I_{eff} = 0.707 \times 30 \text{ amperes}$$
$$I_{eff} = 21.21 \text{ amperes}$$

Figure 1-16 shows the relationship between the various values used to indicate sine-wave amplitude. Review the values in the figure to ensure you understand what each value indicates.

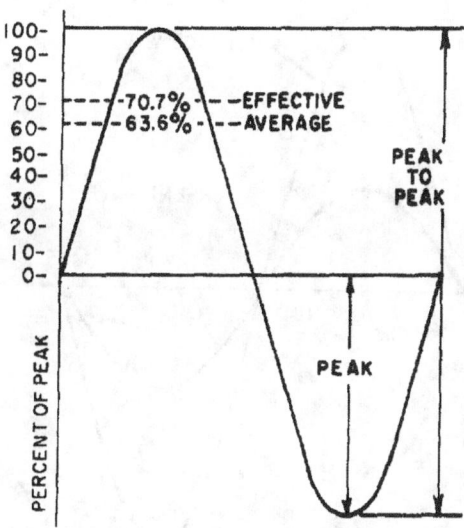

Figure 1-16.—Various values used to indicate sine-wave amplitude.

Q32. What is the most convenient basis for comparing alternating and direct voltages and currents?

Q33. What value of ac is used as a comparison to dc?

Q34. What is the formula for finding the effective value of an alternating current?

Q35. If the peak value of a sine wave is 1,000 volts, what is the effective (E_{eff}) value?

Q36. If I_{eff} = 4.25 ampere, what is I_{max}?

SINE WAVES IN PHASE

When a sine wave of voltage is applied to a resistance, the resulting current is also a sine wave. This follows Ohm's law which states that current is directly proportional to the applied voltage. Now examine figure 1-17. Notice that the sine wave of voltage and the resulting sine wave of current are superimposed on the same time axis. Notice also that as the voltage increases in a positive direction, the current increases along with it, and that when the voltage reverses direction, the current also reverses direction. When two sine waves, such as those represented by figure 1-17, are precisely in step with one another, they are said to be IN PHASE. To be in phase, the two sine waves must go through their maximum and minimum points at the same time and in the same direction.

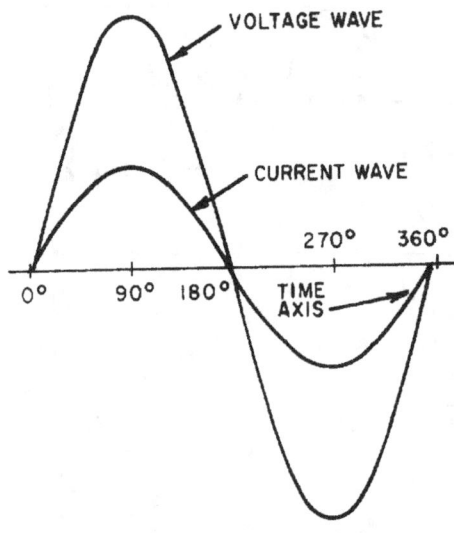

Figure 1-17.—Voltage and current waves in phase.

In some circuits, several sine waves can be in phase with each other. Thus, it is possible to have two or more voltage drops in phase with each other and also be in phase with the circuit current.

SINE WAVES OUT OF PHASE

Figure 1-18 shows voltage wave E_1 which is considered to start at 0° (time one). As voltage wave E_1 reaches its positive peak, voltage wave E_2 starts its rise (time two). Since these voltage waves do not go through their maximum and minimum points at the same instant of time, a PHASE DIFFERENCE exists between the two waves. The two waves are said to be OUT OF PHASE. For the two waves in figure 1-18 the phase difference is 90°.

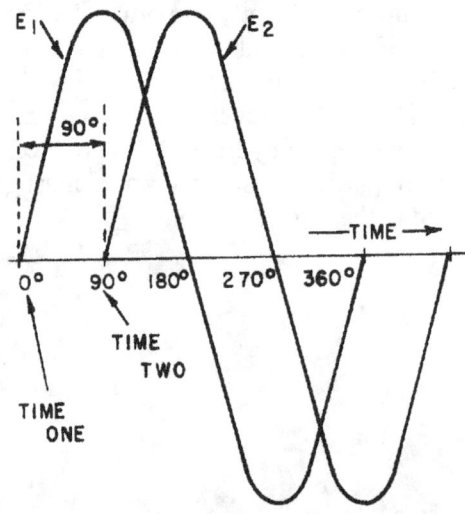

Figure 1-18.—Voltage waves 90° out of phase.

To further describe the phase relationship between two sine waves, the terms LEAD and LAG are used. The amount by which one sine wave leads or lags another sine wave is measured in degrees. Refer again to figure 1-18. Observe that wave E_2 starts 90° later in time than does wave E_1. You can also describe this relationship by saying that wave E_1 leads wave E_2 by 90°, or that wave E_2 lags wave E_1 by 90°. (Either statement is correct; it is the phase relationship between the two sine waves that is important.)

It is possible for one sine wave to lead or lag another sine wave by any number of degrees, except 0° or 360°. When the latter condition exists, the two waves are said to be in phase. Thus, two sine waves that differ in phase by 45° are actually out of phase with each other, whereas two sine waves that differ in phase by 360° are considered to be in phase with each other.

A phase relationship that is quite common is shown in figure 1-19. Notice that the two waves illustrated differ in phase by 180°. Notice also that although the waves pass through their maximum and minimum values at the same time, their instantaneous voltages are always of opposite polarity. If two such waves exist across the same component, and the waves are of equal amplitude, they cancel each other. When they have different amplitudes, the resultant wave has the same polarity as the larger wave and has an amplitude equal to the difference between the amplitudes of the two waves.

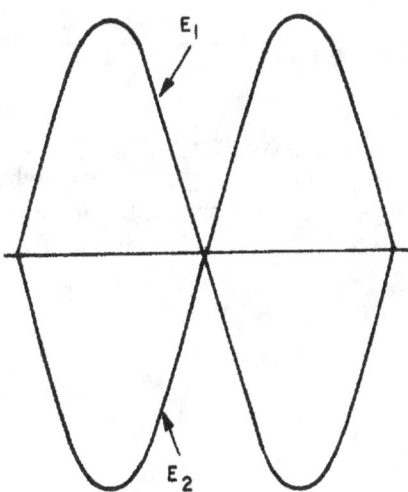

Figure 1-19.—Voltage waves 180° out of phase.

To determine the phase difference between two sine waves, locate the points on the time axis where the two waves cross the time axis traveling in the same direction. The number of degrees between the crossing points is the phase difference. The wave that crosses the axis at the later time (to the right on the time axis) is said to lag the other wave.

Q37. *When are the voltage wave and the current wave in a circuit considered to be in phase?*

Q38. *When are two voltage waves considered to be out of phase?*

Q39. *What is the phase relationship between two voltage waves that differ in phase by 360°?*

Q40. *How do you determine the phase difference between two sine waves that are plotted on the same graph?*

OHM'S LAW IN AC CIRCUITS

Many ac circuits contain resistance only. The rules for these circuits are the same rules that apply to dc circuits. Resistors, lamps, and heating elements are examples of resistive elements. When an ac circuit contains only resistance, Ohm's Law, Kirchhoff's Law, and the various rules that apply to voltage, current, and power in a dc circuit also apply to the ac circuit. The Ohm's Law formula for an ac circuit can be stated as

$$I_{eff} = \frac{E_{eff}}{R} \text{ or } I = \frac{E}{R}$$

Remember, unless otherwise stated, all ac voltage and current values are given as effective values. The formula for Ohm's Law can also be stated as

$$I_{avg} = \frac{E_{avg}}{R} \text{ or } I_{max} = \frac{E_{max}}{R}$$

$$I_{peak-to-peak} = \frac{E_{peak-to-peak}}{R}$$

The important thing to keep in mind is: <u>Do Not mix ac values</u>. When you solve for effective values, all values you use in the formula <u>must be effective values</u>. Similarly, when you solve for average values, all values you use <u>must be average values</u>. This point should be clearer after you work the following problem: A series circuit consists of two resistors (R1 = 5 ohms and R2 = 15 ohms) and an alternating voltage source of 120 volts. What is I_{avg}?

Given: R1 = 5 ohms
R2 = 15 ohms
E_s = 120 ohms

Solution: First solve for total resistance R_T.

$R_T = R1 + R2$
$R_T = 5 \text{ ohms} + 15 \text{ ohms}$
$R_T = 20 \text{ ohms}$

The alternating voltage is assumed to be an effective value (since it is not specified to be otherwise). Apply the Ohm's Law formula.

$$I_{eff} = \frac{E_{eff}}{R}$$

$$I_{eff} = \frac{120 \text{ volts}}{20 \text{ ohms}}$$

$$I_{eff} = 6 \text{ amperes}$$

The problem, however, asked for the average value of current (I_{avg}). To convert the effective value of current to the average value of current, you must first determine the peak or maximum value of current, I_{max}.

$$I_{max} = 1.414 \times I_{eff}$$

$$I_{max} = 1.414 \times 6 \text{ amperes}$$

$$I_{max} = 8.484 \text{ amperes}$$

You can now find I_{avg}. Just substitute 8.484 amperes in the I_{avg} formula and solve for I_{avg}.

$$I_{avg} = 0.636 \times I_{max}$$

$$I_{avg} = 0.636 \times 8.484 \text{ amperes}$$

$$I_{avg} = 5.4 \text{ amperes (rounded off to one decimal place)}$$

Remember, you can use the Ohm's Law formulas to solve any purely resistive ac circuit problem. Use the formulas in the same manner as you would to solve a dc circuit problem.

Q41. *A series circuit consists of three resistors (R1 = 10Ω, R2 = 20Ω, R3 = 15Ω) and an alternating voltage source of 100 volts. What is the effective value of current in the circuit?*

Q42. *If the alternating source in Q41 is changed to 200 volts peak-to-peak, what is I_{avg}?*

Q43. *If E_{eff} is 130 volts and I_{eff} is 3 amperes, what is the total resistance (R_T) in the circuit?*

SUMMARY

Before going on to chapter 2, read the following summary of the material in chapter 1. This summary will reinforce what you have already learned.

DC AND AC—Direct current flows in one direction only, while alternating current is constantly changing in amplitude and direction.

ADVANTAGES AND DISADVANTAGES OF AC AND DC—Direct current has several disadvantages compared to alternating current. Direct current, for example, must be generated at the voltage level required by the load. Alternating current, however, can be generated at a high level and stepped down at the consumer end (through the use of a transformer) to whatever voltage level is required by the load. Since power in a dc system must be transmitted at low voltage and high current levels, the I^2R power loss becomes a problem in the dc system. Since power in an ac system can be transmitted at a high voltage level and a low current level, the I^2R power loss in the ac system is much less than that in the dc system.

VOLTAGE WAVEFORMS—The waveform of voltage or current is a graphical picture of changes in voltage or current values over a period of time.

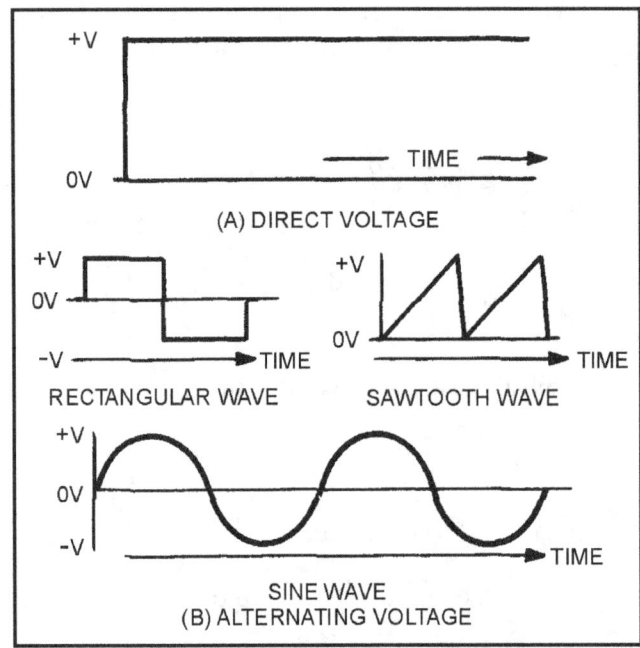

ELECTROMAGNETISM—When a compass is placed in the vicinity of a current-carrying conductor, the needle aligns itself at right angles to the conductor. The north pole of the compass indicates the direction of the magnetic field produced by the current. By knowing the direction of current, you can use the left-hand rule for conductors to determine the direction of the magnetic lines of force.

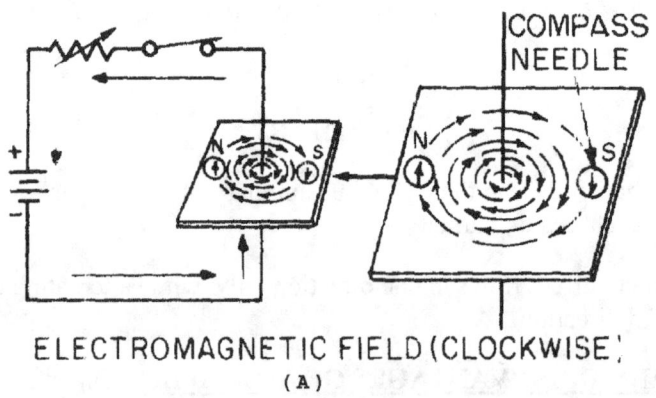

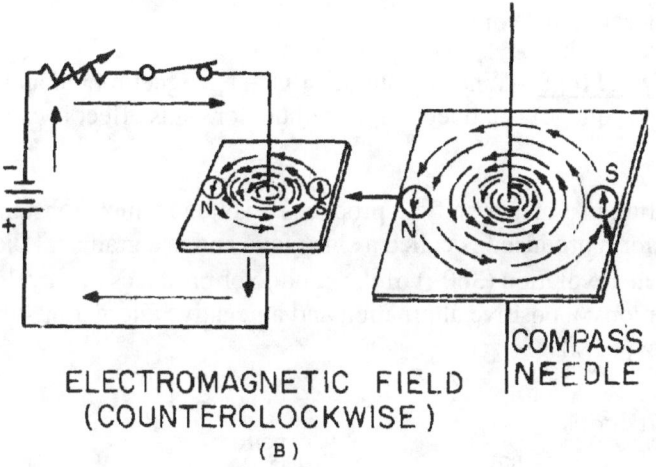

**ELECTROMAGNETIC FIELD
(COUNTERCLOCKWISE)**
(B)

Arrows are generally used in electrical diagrams to indicate the direction of current in a wire. A cross (+) on the end of a cross-sectional view of a wire indicates that current is flowing away from you, while a dot (·) indicates that current is flowing toward you.

When two adjacent parallel conductors carry current in the same direction, the magnetic fields around the conductors aid each other. When the currents in the two conductors flow in opposite directions, the fields around the conductors oppose each other.

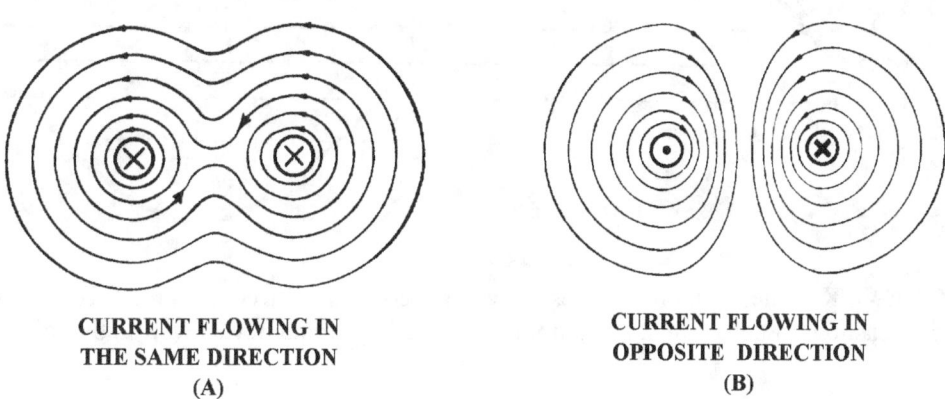

**CURRENT FLOWING IN
THE SAME DIRECTION**
(A)

**CURRENT FLOWING IN
OPPOSITE DIRECTION**
(B)

MAGNETIC FIELD OF A COIL—When wire is wound around a core, it forms a COIL. The magnetic fields produced when current flows in the coil combine. The combined influence of all of the fields around the turns produce a two-pole field similar to that of a simple bar magnet.

When the direction of current in the coil is reversed, the polarity of the two-pole field of the coil is reversed.

The strength of the magnetic field of the coil is dependent upon:

- The number of turns of the wire in the coil.

- The amount of current in the coil.

- The ratio of the coil length to the coil width.

- The type of material in the core.

BASIC AC GENERATION—When a conductor is in a magnetic field and either the field or the conductor moves, an emf (voltage) is induced in the conductor. This effect is called electromagnetic induction.

A loop of wire rotating in a magnetic field produces a voltage which constantly changes in amplitude and direction. The waveform produced is called a sine wave and is a graphical picture of alternating current (ac). One complete revolution (360°) of the conductor produces one cycle of ac. The cycle is composed of two alternations: a positive alternation and a negative alternation. One cycle of ac in one second is equal to 1 hertz (1 Hz).

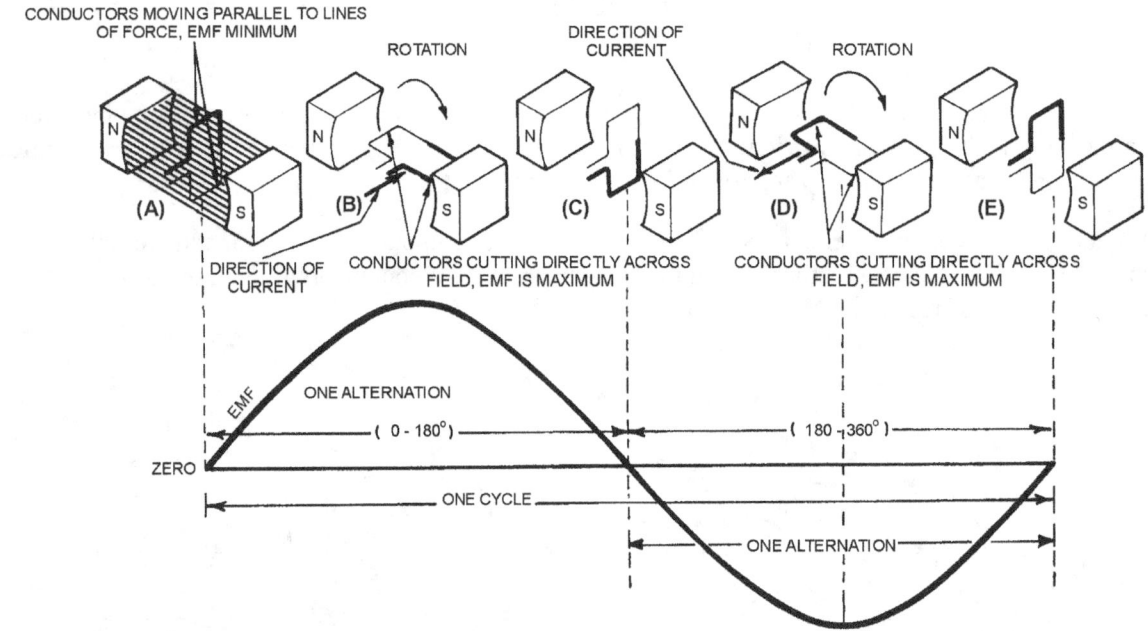

FREQUENCY—The number of cycles of ac per second is referred to as the FREQUENCY. AC frequency is measured in hertz. Most ac equipment is rated by frequency as well as by voltage and current.

PERIOD—The time required to complete one cycle of a waveform is called the PERIOD OF THE WAVE.

Each ac sine wave is composed of two alternations. The alternation which occurs during the time the sine wave is positive is called the positive alternation. The alternation which occurs during the time the sine wave is negative is called the negative alternation. In each cycle of sine wave, the two alternations are identical in size and shape, but opposite in polarity.

The period of a sine wave is inversely proportional to the frequency; e.g., the higher the frequency, the shorter the period. The mathematical relationships between time and frequency are

$$t = \frac{1}{f} \text{ and } f = \frac{1}{t}$$

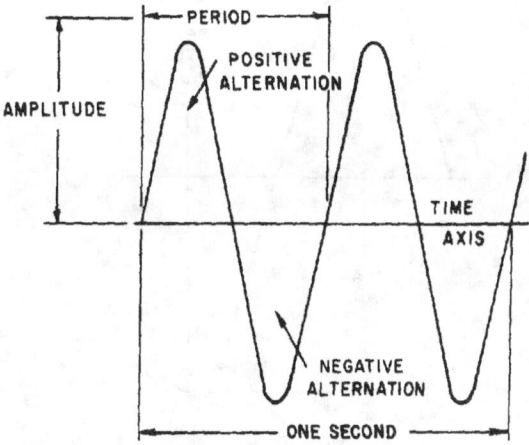

WAVELENGTH—The period of a sine wave is defined as the time it takes to complete one cycle. The distance the waveform covers during this period is referred to as the wavelength. Wavelength is indicated by lambda (λ) and is measured from a point on a given waveform (sine wave) to the corresponding point on the next waveform.

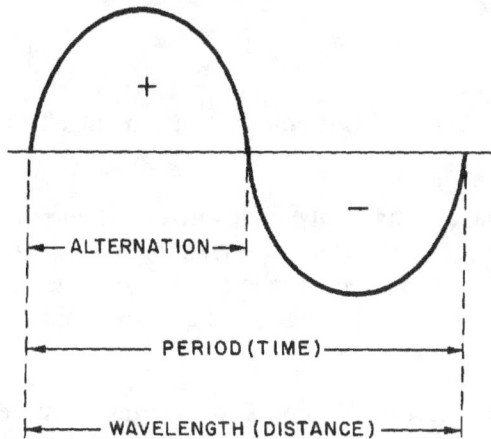

PEAK AND PEAK-TO-PEAK VALUES—The maximum value reached during one alternation of a sine wave is the peak value. The maximum reached during the positive alternation to the maximum value reached during the negative alternation is the peak-to-peak value. The peak-to-peak value is twice the peak value.

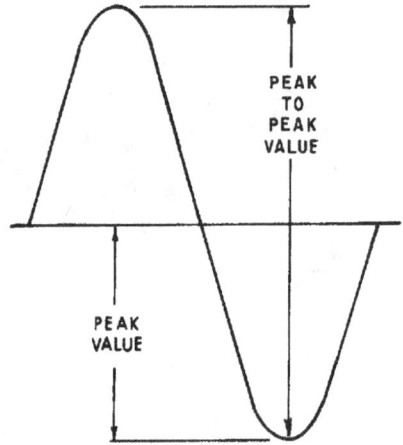

INSTANTANEOUS VALUE—The instantaneous value of a sine wave of alternating voltage or current is the value of voltage or current at one particular instant of time. There are an infinite number of instantaneous values between zero and the peak value.

AVERAGE VALUE—The average value of a sine wave of voltage or current is the average of all the instantaneous values during one alternation. The average value is equal to 0.636 of the peak value. The formulas for average voltage and average current are:

$$E_{avg} = 0.636 \times E_{max}$$
$$I_{avg} = 0.636 \times I_{max}$$

Remember: The average value (E_{avg} or I_{avg}) is for one alternation only. The average value of a complete sine wave is zero.

EFFECTIVE VALUE—The effective value of an alternating current or voltage is the value of alternating current or voltage that produces the same amount of heat in a resistive component that would be produced in the same component by a direct current or voltage of the same value. The effective value of a sine wave is equal to 0.707 times the peak value. The effective value is also called the root mean square or rms value.

The term rms value is used to describe the process of determining the effective value of a sine wave by using the instantaneous value of voltage or current. You can find the rms value of a current or voltage by taking equally spaced instantaneous values on the sine wave and extracting the square root of the average of the sum of the instantaneous values. This is where the term "Root-Mean-Square" (rms) value comes from.

The formulas for effective and maximum values of voltage and current are:

$$E_{eff} = 0.707 \times E_{max}$$
$$E_{max} = 1.414 \times E_{eff}$$
$$I_{eff} = 0.707 \times I_{max}$$
$$I_{max} = 1.414 \times I_{eff}$$

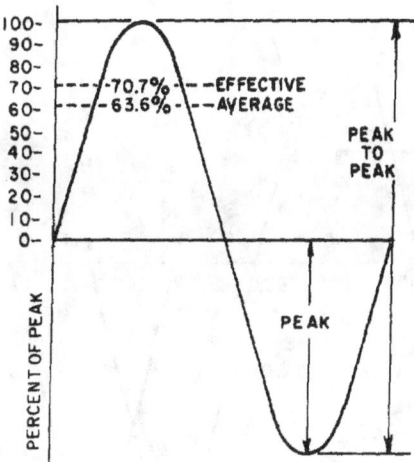

SINE WAVES IN PHASE—When two sine waves are exactly in step with each other, they are said to be in phase. To be in phase, both sine waves must go through their minimum and maximum points at the same time and in the same direction.

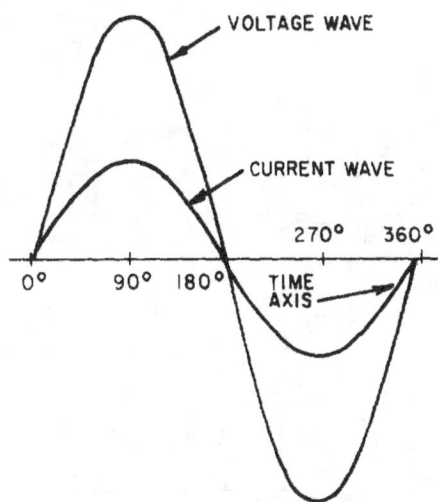

SINE WAVES OUT OF PHASE—When two sine waves go through their minimum and maximum points at different times, a phase difference exists between them. The two waves are said to be out of phase with each other. To describe this phase difference, the terms lead and lag are used. The wave that reaches its minimum (or maximum) value first is said to lead the other wave. The term lag is used to describe the wave that reaches its minimum (or maximum) value some time after the first wave does. When a sine wave is described as leading or lagging, the difference in degrees is usually stated. For example, wave E_1 leads wave E_2 by 90°, or wave E_2 lags wave E_1 by 90°. Remember: Two sine waves can differ by any number of degrees except 0° and 360°. Two sine waves that differ by 0° or by 360° are considered to be in phase. Two sine waves that are opposite in polarity and that differ by 180° are said to be out of phase, even though they go through their minimum and maximum points at the same time.

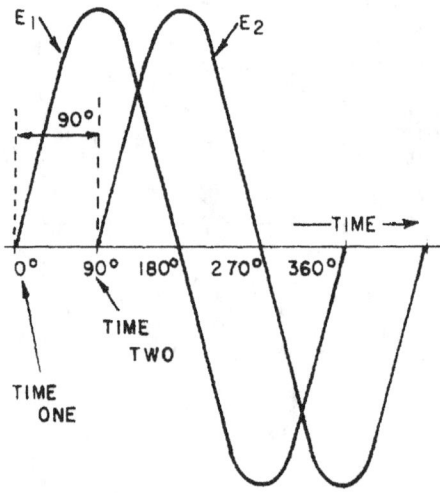

OHM'S LAW IN AC CIRCUIT—All dc rules and laws apply to an ac circuit that contains only resistance. The important point to remember is: Do not mix ac values. Ohm's Law formulas for ac circuits are given below:

$$I = \frac{E}{R}$$

$$I_{eff} = \frac{E_{eff}}{R}$$

$$I_{avg} = \frac{E_{avg}}{R}$$

$$I_{max} = \frac{E_{max}}{R}$$

$$I_{Peak-to-Peak} = \frac{E_{Peak-to-Peak}}{R}$$

ANSWERS TO QUESTIONS Q1. THROUGH Q43.

A1. An electrical current which flows in one direction only.

A2. An electrical current which is constantly varying in amplitude, and which changes direction at regular intervals.

A3. The dc voltage must be generated at the level required by the load.

A4. The I^2R power loss is excessive.

A5. Alternating current (ac).

A6. The needle aligns itself at right angles to the conductor.

A7. (a) clockwise (b) counterclockwise.

A8. It is used to determine the relation between the direction of the magnetic lines of force around a conductor and the direction of current through the conductor.

A9. The north pole of the compass will point in the direction of the magnetic lines of force.

A10. It combines with the other field.

A11. It deforms the other field.

A12. (a) The field consists of concentric circles in a plane perpendicular to the wire (b) the field of each turn of wire links with the fields of adjacent turns producing a two-pole field similar in shape to that of a simple bar magnet.

A13. The polarity of the two-pole field reverses.

A14. Use the left-hand rule for coils.

A15. Grasp the coil in your left hand, with your fingers "wrapped around" in the direction of electron flow. The thumb will point toward the north pole.

A16. (a) When the conductors are cutting directly across the magnetic lines of force (at the 90° and 270° points). (b) When the conductors are moving parallel to the magnetic lines of force (at the 0°, 180°, and 360° points).

A17. 360°.

A18. Extend your left hand so that your thumb points in the direction of conductor movement, and your forefinger points in the direction of the magnetic flux (north to south). Now point your middle finger 90° from the forefinger and it will point in the direction of electron current flow in the conductor.

A19. Continuous rotation of the conductor through magnetic fines of force produces a series of cycles of alternating voltage or, in other words, an alternating voltage or a sine wave of voltage.

A20. Frequency is the number of complete cycles of alternating voltage or current completed each second.

A21. Period.

A22. A positive alternation is the positive variation in the voltage or current of a sine curve.

A23. The period measures time and the wavelength measures distance.

A24. The peak value is the maximum value of one alternation; the peak-to-peak value is twice the maximum or peak value.

A25. Twice.

A26. The instantaneous value (E_{inst} or I_{inst})

A27. Average value (E_{avg} or I_{avg})

A28. Zero

A29.
$$I_{avg} = 0.636 \times I_{max}$$
$$E_{avg} = 0.636 \times E_{max}$$

A30.
$$E_{avg} = 0.636 \times 115 \text{ volts}$$
$$E_{avg} = 73.14 \text{ volts}$$

A31.
$$\text{If } I_{avg} = I_{max} \times 0.636, \text{ then } I_{max} = \frac{I_{avg}}{0.636}$$
Thus,
$$I_{max} = \frac{1.272}{0.636} \text{ ampere} = 2 \text{ amperes}$$

A32. The power (heat) produced in a resistance by a dc voltage is compared to that produced in the same resistance by an ac voltage of the same peak amplitude.

A33. The effective value.

A34.
$$I_{eff} = 0.707 \times I_{max}$$

A35.
$$E_{eff} = 0.707 \times E_{max}$$
$$= 0.707 \times E_{max}$$
$$= 0.707 \times 1,000 \text{ volts}$$
$$E_{eff} = 707 \text{ volts}$$

A36.
$$I_{max} = 1.414 \times I_{eff}$$
$$= 1.414 \times 4.25 \text{ amperes}$$
$$= 6 \text{ amperes.}$$

(Remember: Unless specified otherwise, the voltage or current value is always considered to be the effective value.)

A37. When the two waves go through their maximum and minimum points at the same time and in the same direction.

A38. When the waves do not go through their maximum and minimum points at the same time, a PHASE DIFFERENCE exists, and the two waves are said to be out of phase. (Two waves are also considered to be out of phase if they differ in phase by 180° and their instantaneous voltages are always of opposite polarity, even though both waves go through their maximum and minimum points at the same time).

A39. They are in phase with each other.

A40. Locate the points on the time axis where the two waves cross traveling in the same direction. The number of degrees between these two points is the phase difference.

A41.
$$I_{eff} = \frac{100}{45} = 2.22 \text{ ampers}$$

A42. $I_{avg} = 0.636 \times I_{max} = 1.41$ amperes.

A43. 43.3 ohms.

CHAPTER 2
INDUCTANCE

LEARNING OBJECTIVES

Upon completion of this chapter you will be able to:

1. Write the basic unit of and the symbol for inductance.

2. State the type of moving field used to generate an emf in a conductor.

3. Define the term "inductance."

4. State the meanings of the terms "induced emf" and "counter emf."

5. State Lenz's law.

6. State the effect that inductance has on steady direct current, and direct current that is changing in magnitude.

7. List five factors that affect the inductance of a coil, and state how various physical changes in these factors affect inductance.

8. State the principles and sequences involved in the buildup and decay of current in an LR series circuit.

9. Write the formula for computing one time constant in an LR series circuit.

10. Solve L/R time constant problems.

11. State the three types of power loss in an inductor.

12. Define the term "mutual inductance."

13. State the meaning of the term "coupled circuits."

14. State the meaning of the term "coefficient of coupling."

15. Given the inductance values of and the coefficient of coupling between two series-connected inductors, solve for mutual inductance, M.

16. Write the formula for the "total inductance" of two inductors connected in series-opposing.

17. Given the inductance values of and the mutual inductance value between two coils connected in series-aiding, solve for their combined inductance, L_T.

INDUCTANCE

The study of inductance presents a very challenging but rewarding segment of electricity. It is challenging in the sense that, at first, it will seem that new concepts are being introduced. You will realize as this chapter progresses that these "new concepts" are merely extensions and enlargements of fundamental principles that you learned previously in the study of magnetism and electron physics. The study of inductance is rewarding in the sense that a thorough understanding of it will enable you to acquire a working knowledge of electrical circuits more rapidly.

CHARACTERISTICS OF INDUCTANCE

Inductance is the characteristic of an electrical circuit that opposes the starting, stopping, or a change in value of current. The above statement is of such importance to the study of inductance that it bears repeating. Inductance is the characteristic of an electrical conductor that OPPOSES CHANGE in CURRENT. The symbol for inductance is L and the basic unit of inductance is the HENRY (H). One henry is equal to the inductance required to induce one volt in an inductor by a change of current of one ampere per second.

You do not have to look far to find a physical analogy of inductance. Anyone who has ever had to push a heavy load (wheelbarrow, car, etc.) is aware that it takes more work to start the load moving than it does to keep it moving. Once the load is moving, it is easier to keep the load moving than to stop it again. This is because the load possesses the property of INERTIA. Inertia is the characteristic of mass which opposes a CHANGE in velocity. Inductance has the same effect on current in an electrical circuit as inertia has on the movement of a mechanical object. It requires more energy to start or stop current than it does to keep it flowing.

Q1. What is the basic unit of inductance and the abbreviation for this unit?

ELECTROMOTIVE FORCE (EMF)

You have learned that an electromotive force is developed whenever there is <u>relative motion</u> between a magnetic field and a conductor.

Electromotive force is a difference of potential or voltage which exists between two points in an electrical circuit. In generators and inductors the emf is developed by the action between the magnetic field and the electrons in a conductor. This is shown in figure 2-1.

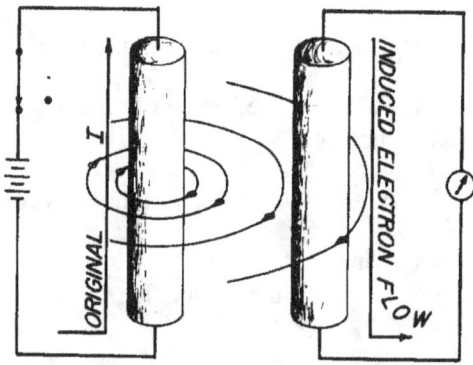

Figure 2-1.—Generation of an emf in an electrical conductor.

When a magnetic field moves through a stationary metallic conductor, electrons are dislodged from their orbits. The electrons move in a direction determined by the movement of the magnetic lines of flux. This is shown below:

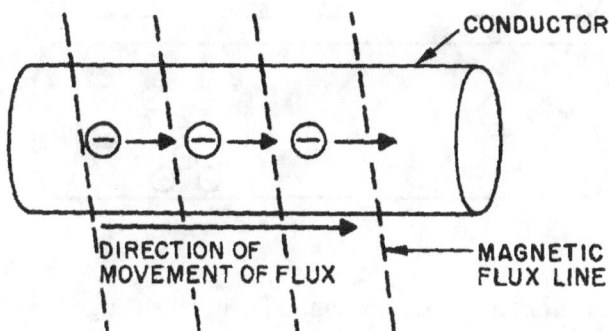

The electrons move from one area of the conductor into another area. The area that the electrons moved from has fewer negative charges (electrons) and becomes positively charged. The area the electrons move into becomes negatively charged. This is shown below:

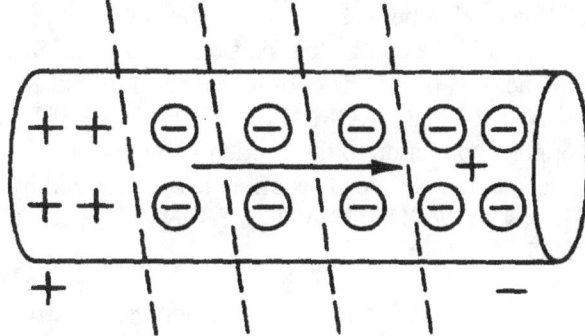

The difference between the charges in the conductor is equal to a difference of potential (or voltage). This voltage caused by the moving magnetic field is called electromotive force (emf).

In simple terms, the action of a moving magnetic field on a conductor can be compared to the action of a broom. Consider the moving magnetic field to be a moving broom. As the magnetic broom moves along (through) the conductor, it gathers up and pushes electrons before it, as shown below:

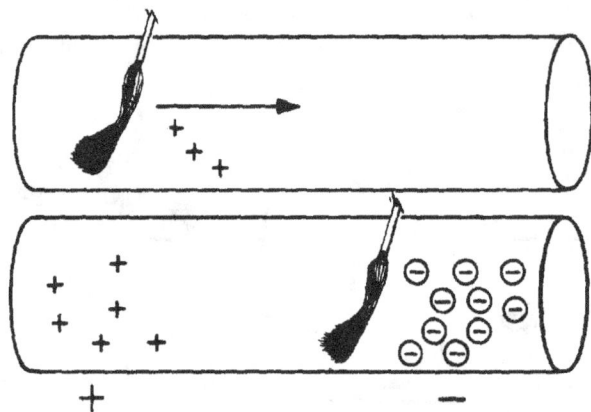

The area from which electrons are moved becomes positively charged, while the area into which electrons are moved becomes negatively charged. The potential difference between these two areas is the electromotive force or emf.

Q2. An emf is generated in a conductor when the conductor is cut by what type of field?

SELF-INDUCTANCE

Even a perfectly straight length of conductor has some inductance. As you know, current in a conductor produces a magnetic field surrounding the conductor. When the current changes, the magnetic field changes. This causes relative motion between the magnetic field and the conductor, and an electromotive force (emf) is induced in the conductor. This emf is called a SELF-INDUCED EMF because it is induced in the conductor carrying the current. The emf produced by this moving magnetic field is also referred to as COUNTER ELECTROMOTIVE FORCE (cemf). The polarity of the counter electromotive force is in the opposite direction to the applied voltage of the conductor. The overall effect will be to oppose a change in current magnitude. This effect is summarized by Lenz's law which states that: THE INDUCED EMF IN ANY CIRCUIT IS ALWAYS IN A DIRECTION TO OPPOSE THE EFFECT THAT PRODUCED IT.

If the shape of the conductor is changed to form a loop, then the electromagnetic field around each portion of the conductor cuts across some other portion of the same conductor. This is shown in its simplest form in figure 2-2. A length of conductor is looped so that two portions of the conductor lie next to each other. These portions are labeled conductor 1 and conductor 2. When the switch is closed, current (electron flow) in the conductor produces a magnetic field around ALL portions of the conductor. For simplicity, the magnetic field (expanding lines of flux) is shown in a single plane that is perpendicular to both conductors. Although the expanding field of flux originates at the same time in both conductors, it is considered as originating in conductor 1 and its effect on conductor 2 will be explained. With increasing current, the flux field expands outward from conductor 1, cutting across a portion of conductor 2. This results in an induced emf in conductor 2 as shown by the dashed arrow. Note that the induced emf is in the opposite direction to (in OPPOSITION to) the battery current and voltage, as stated in Lenz's law.

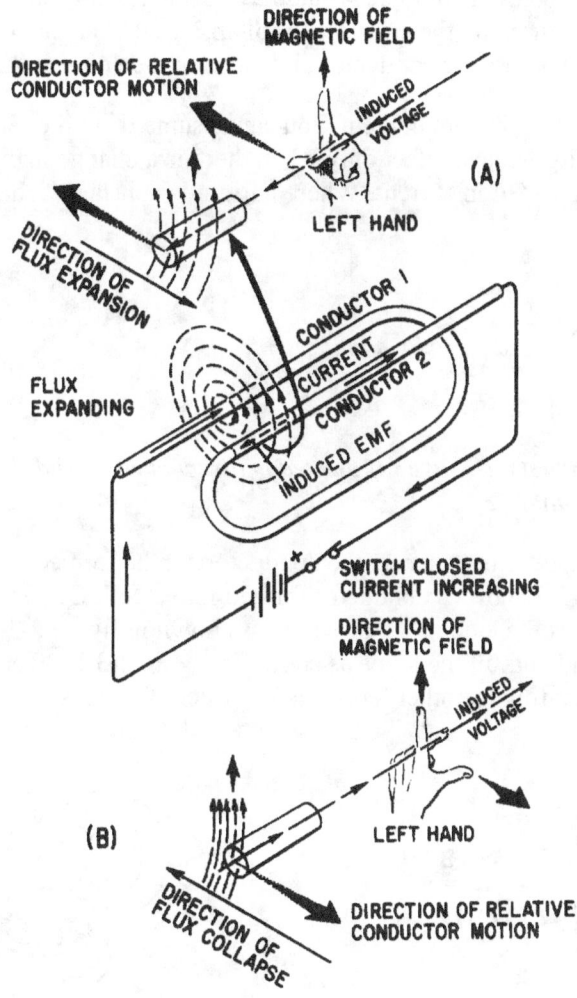

Figure 2-2.—Self-inductance.

The direction of this induced voltage may be determined by applying the LEFT-HAND RULE FOR GENERATORS. This rule is applied to a portion of conductor 2 that is "lifted" and enlarged for this purpose in figure 2-2(A). This rule states that if you point the thumb of your left hand in the direction of relative motion of the conductor and your index finger in the direction of the magnetic field, your middle finger, extended as shown, will now indicate the direction of the induced current which will generate the induced voltage (cemf) as shown.

In figure 2-2(B), the same section of conductor 2 is shown after the switch has been opened. The flux field is collapsing. Applying the left-hand rule in this case shows that the reversal of flux MOVEMENT has caused a reversal in the direction of the induced voltage. The induced voltage is now in the same direction as the battery voltage. The most important thing for you to note is that the self-induced voltage opposes BOTH changes in current. That is, when the switch is closed, this voltage delays the initial buildup of current by opposing the battery voltage. When the switch is opened, it keeps the current flowing in the same direction by aiding the battery voltage.

Thus, from the above explanation, you can see that when a current is building up it produces an expanding magnetic field. This field induces an emf in the direction opposite to the actual flow of current.

This induced emf opposes the growth of the current and the growth of the magnetic field. If the increasing current had not set up a magnetic field, there would have been no opposition to its growth. The whole reaction, or opposition, is caused by the creation or collapse of the magnetic field, the lines of which as they expand or contract cut across the conductor and develop the counter emf.

Since all circuits have conductors in them, you can assume that all circuits have inductance. However, inductance has its greatest effect only when there is a change in current. Inductance does NOT oppose current, only a CHANGE in current. Where current is constantly changing as in an ac circuit, inductance has more effect.

Q3. *Define inductance.*

Q4. *What is meant by induced emf? By counter emf?*

Q5. *State Lenz's law.*

Q6. *What effect does inductance have (a) on steady direct current and (b) on direct current while it is changing in amplitude?*

To increase the property of inductance, the conductor can be formed into a loop or coil. A coil is also called an inductor. Figure 2-3 shows a conductor formed into a coil. Current through one loop produces a magnetic field that encircles the loop in the direction as shown in figure 2-3(A). As current increases, the magnetic field expands and cuts all the loops as shown in figure 2-3(B). The current in each loop affects all other loops. The field cutting the other loop has the effect of increasing the opposition to a current change.

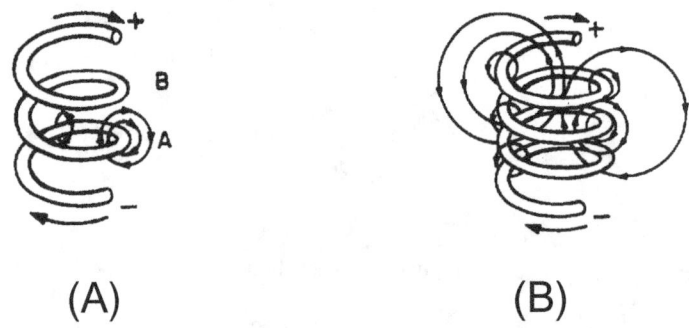

Figure 2-3.—Inductance.

Inductors are classified according to core type. The core is the center of the inductor just as the core of an apple is the center of an apple. The inductor is made by forming a coil of wire around a core. The core material is normally one of two basic types: soft-iron or air. An iron-core inductor and its schematic symbol (which is represented with lines across the top of it to indicate the presence of an iron core) are shown, in figure 2-4(A). The air-core inductor may be nothing more than a coil of wire, but it is usually a coil formed around a hollow form of some nonmagnetic material such as cardboard. This material serves no purpose other than to hold the shape of the coil. An air-core inductor and its schematic symbol are shown in figure 2-4(B).

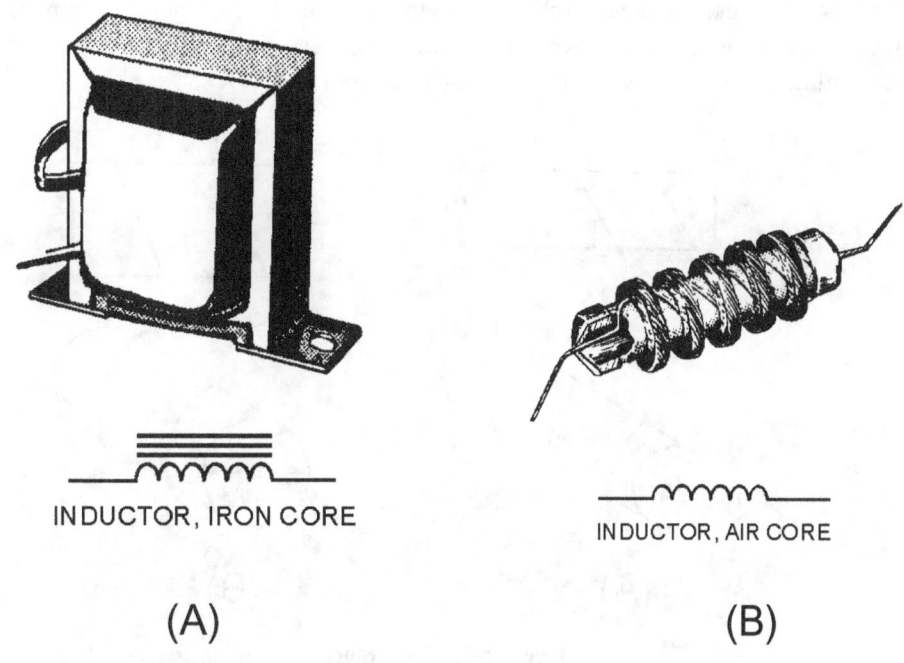

Figure 2-4.—Inductor types and schematic symbols.

Factors Affecting Coil Inductance

There are several physical factors which affect the inductance of a coil. They include the number of turns in the coil, the diameter of the coil, the coil length, the type of material used in the core, and the number of layers of winding in the coils.

Inductance depends entirely upon the physical construction of the circuit, and can only be measured with special laboratory instruments. Of the factors mentioned, consider first how the number of turns affects the inductance of a coil. Figure 2-5 shows two coils. Coil (A) has two turns and coil (B) has four turns. In coil (A), the flux field set up by one loop cuts one other loop. In coil (B), the flux field set up by one loop cuts three other loops. Doubling the number of turns in the coil will produce a field twice as strong, if the same current is used. A field twice as strong, cutting twice the number of turns, will induce four times the voltage. Therefore, it can be said that the inductance varies as the square of the number of turns.

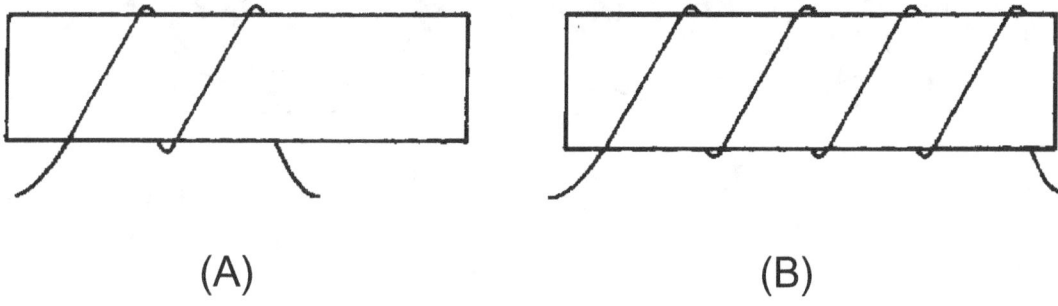

Figure 2-5.—Inductance factor (turns).

The second factor is the coil diameter. In figure 2-6 you can see that the coil in view B has twice the diameter of coil view A. Physically, it requires more wire to construct a coil of large diameter than one of small diameter with an equal number of turns. Therefore, more lines of force exist to induce a counter emf in the coil with the larger diameter. Actually, the inductance of a coil <u>increases directly as the cross-sectional area of the core increases</u>. Recall the formula for the area of a circle: $A = \pi r^2$. Doubling the radius of a coil increases the inductance by a factor of four.

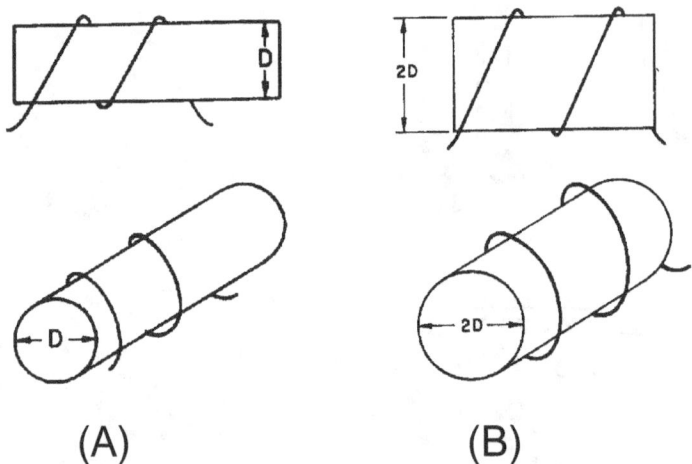

Figure 2-6.—Inductance factor (diameter).

The third factor that affects the inductance of a coil is the length of the coil. Figure 2-7 shows two examples of coil spacings. Coil (A) has three turns, rather widely spaced, making a relatively long coil. A coil of this type has few flux linkages, due to the greater distance between each turn. Therefore, coil (A) has a relatively low inductance. Coil (B) has closely spaced turns, making a relatively short coil. This close spacing increases the flux linkage, increasing the inductance of the coil. <u>Doubling the length of a coil while keeping the same number of turns halves the value of inductance.</u>

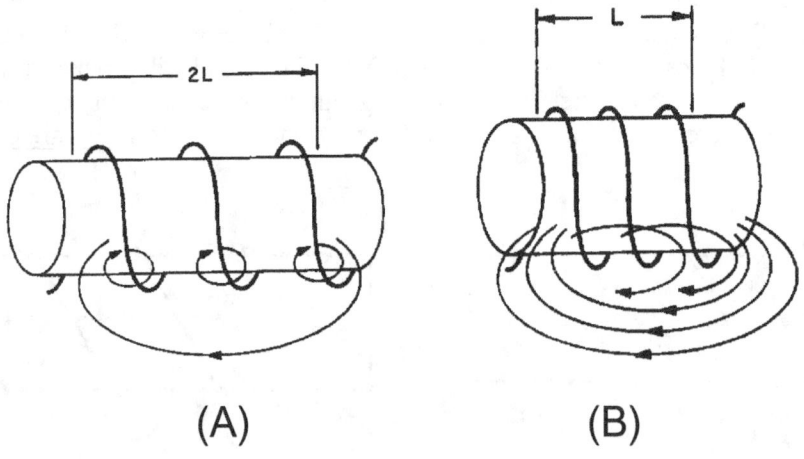

Figure 2-7.—Inductance factor (coil length). CLOSELY WOUND

The fourth physical factor is the type of core material used with the coil. Figure 2-8 shows two coils: Coil (A) with an air core, and coil (B) with a soft-iron core. The magnetic core of coil (B) is a better path for magnetic lines of force than is the nonmagnetic core of coil (A). The soft-iron magnetic core's high permeability has less reluctance to the magnetic flux, resulting in more magnetic lines of force. This increase in the magnetic lines of force increases the number of lines of force cutting each loop of the coil, thus increasing the inductance of the coil. It should now be apparent that the <u>inductance of a coil increases directly as the permeability of the core material increases</u>.

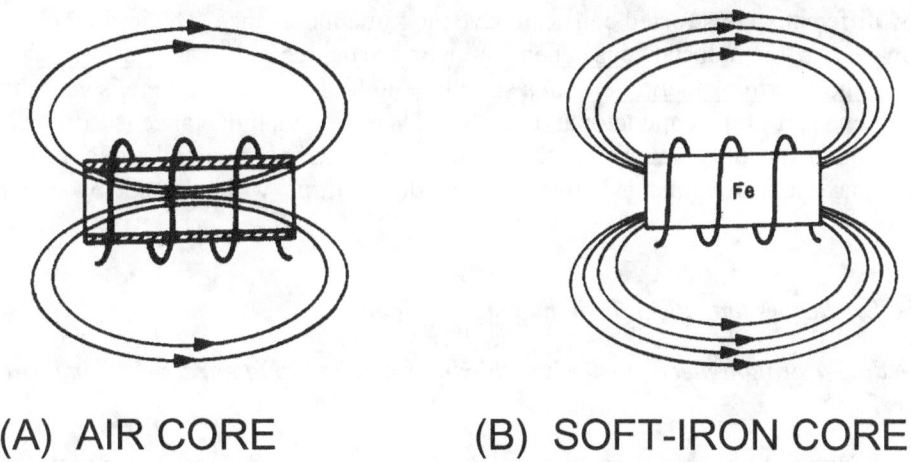

(A) AIR CORE   (B) SOFT-IRON CORE

Figure 2-8.—Inductance factor (core material).

Another way of increasing the inductance is to wind the coil in layers. Figure 2-9 shows three cores with different amounts of layering. The coil in figure 2-9(A) is a poor inductor compared to the others in the figure because its turns are widely spaced and there is no layering. The flux movement, indicated by the dashed arrows, does not link effectively because there is only one layer of turns. A more inductive coil is shown in figure 2-9(B). The turns are closely spaced and the wire has been wound in two layers. The two layers link each other with a greater number of flux loops during all flux movements. Note that nearly all the turns, such as X, are next to four other turns (shaded). This causes the flux linkage to be increased.

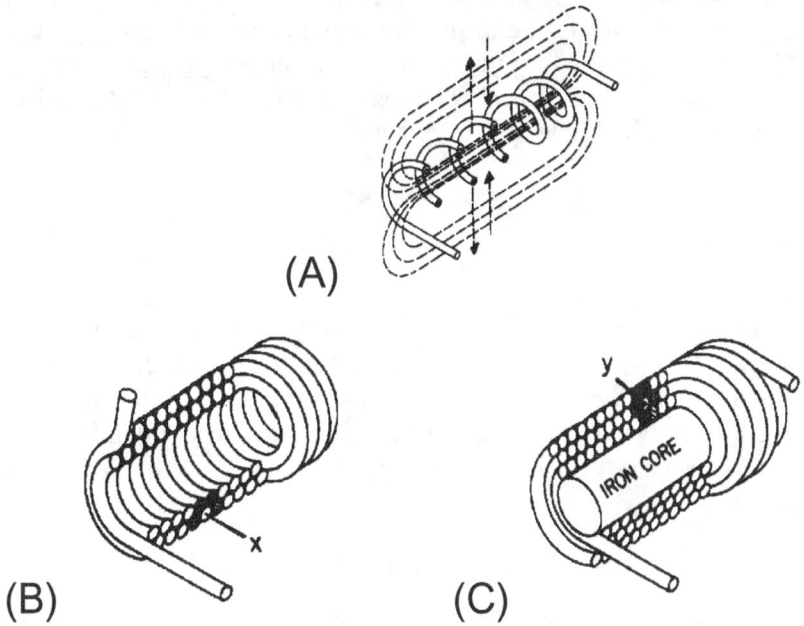

Figure 2-9.—Coils of various inductances.

2-9

A coil can be made still more inductive by winding it in three layers, as shown in figure 2-9(C). The increased number of layers (cross-sectional area) improves flux linkage even more. Note that some turns, such as Y, lie directly next to six other turns (shaded). In actual practice, layering can continue on through many more layers. The important fact to remember, however, is that <u>the inductance of the coil increases with each layer added</u>.

As you have seen, several factors can affect the inductance of a coil, and all of these factors are variable. Many differently constructed coils can have the same inductance. The important information to remember, however, is that <u>inductance is dependent upon the degree of linkage between the wire conductor(s) and the electromagnetic field</u>. In a straight length of conductor, there is very little flux linkage between one part of the conductor and another. Therefore, its inductance is extremely small. It was shown that conductors become much more inductive when they are wound into coils. This is true because there is maximum flux linkage between the conductor turns, which lie side by side in the coil.

Q7.

    a. *List five factors that affect the inductance of a coil.*

    b. *Bending a straight piece of wire into a loop or coil has what effect on the inductance of the wire?*

    c. *Doubling the number of turns in a coil has what effect on the inductance of the coil?*

    d. *Decreasing the diameter of a coil has what effect on the inductance of the coil?*

    e. *Inserting a soft-iron core into a coil has what effect on the inductance of the coil?*

    f. *Increasing the number of layers of windings in a coil has what effect on the inductance of the coil?*

UNIT OF INDUCTANCE

As stated before, the basic unit of inductance (L) is the HENRY (H), named after Joseph Henry, the co-discoverer with Faraday of the principle of electromagnetic induction. <u>An inductor has an inductance of 1 henry if an emf of 1 volt is induced in the inductor when the current through the inductor is changing at the rate of 1 ampere per second</u>. The relationship between the induced voltage, the inductance, and the rate of change of current with respect to time is stated mathematically as:

$$E_{ind} = L \frac{\Delta I}{\Delta t}$$

where E_{ind} is the induced emf in volts; L is the inductance in henrys; and ΔI is the change in current in amperes occurring in Δt seconds. The symbol Δ (Greek letter delta), means "a change in ....". The henry is a large unit of inductance and is used with relatively large inductors. With small inductors, the millihenry is used. (A millihenry is equal to 1×10^{-3} henry, and one henry is equal to 1,000 millihenrys.) For still smaller inductors the unit of inductance is the microhenry (μH). (A $\mu H = 1 \times 10^{-6} H$, and one henry is equal to 1,000,000 microhenrys.)

GROWTH AND DECAY OF CURRENT IN AN LR SERIES CIRCUIT

When a battery is connected across a "pure" inductance, the current builds up to its final value at a rate determined by the battery voltage and the internal resistance of the battery. The current buildup is gradual because of the counter emf generated by the self-inductance of the coil. When the current starts to flow, the magnetic lines of force move outward from the coil. These lines cut the turns of wire on the inductor and build up a counter emf that opposes the emf of the battery. This opposition causes a delay in the time it takes the current to build up to a steady value. When the battery is disconnected, the lines of force collapse. Again these lines cut the turns of the inductor and build up an emf that tends to prolong the flow of current.

A voltage divider containing resistance and inductance may be connected in a circuit by means of a special switch, as shown in figure 2-10(A). Such a series arrangement is called an LR series circuit.

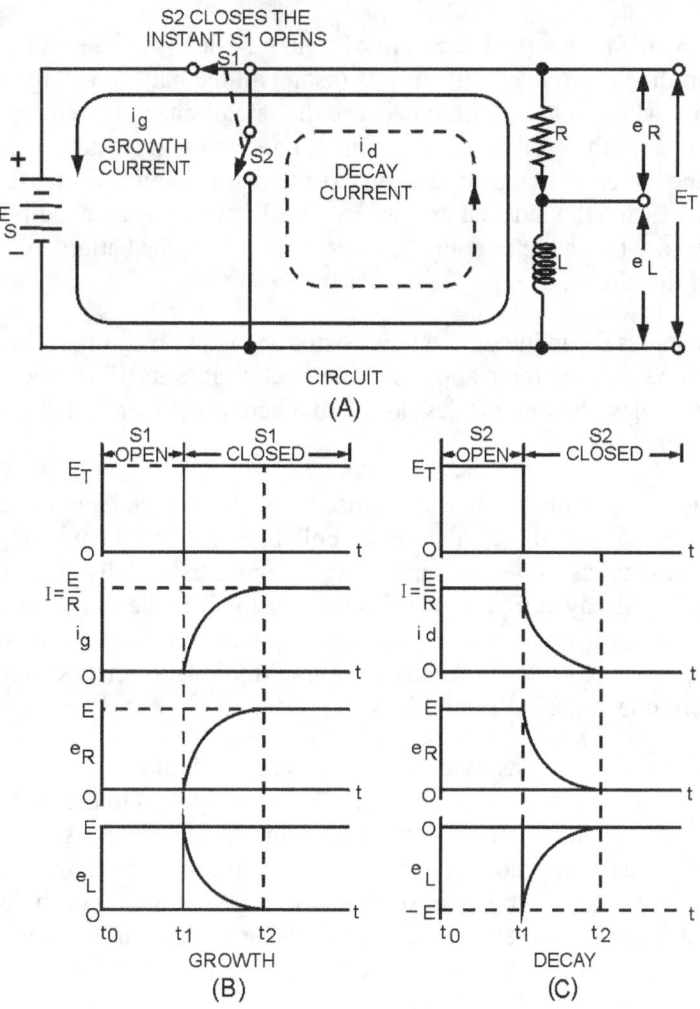

Figure 2-10.—Growth and decay of current in an LR series circuit.

When switch S_1 is closed (as shown), a voltage E_S appears across the voltage divider. At this instant the current will attempt to increase to its maximum value. However, this instantaneous current change causes coil L to produce a back EMF, which is opposite in polarity and almost equal to the EMF of the source. This back EMF opposes the rapid current change. Figure 2-10(B) shows that at the instant switch S_1 is closed, there is no measurable growth current (i_g), a minimum voltage drop is across resistor R, and maximum voltage exists across inductor L.

As current starts to flow, a voltage (e_R) appears across R, and the voltage across the inductor is reduced by the same amount. The fact that the voltage across the inductor (L) is reduced means that the growth current (i_g) is increased and consequently e_R is increased. Figure 2-10(B) shows that the voltage across the inductor (e_L) finally becomes zero when the growth current (i_g) stops increasing, while the voltage across the resistor (e_R) builds up to a value equal to the source voltage (E_S).

Electrical inductance is like mechanical inertia, and the growth of current in an inductive circuit can be likened to the acceleration of a boat on the surface of the water. The boat does not move at the instant a constant force is applied to it. At this instant all the applied force is used to overcome the inertia of the boat. Once the inertia is overcome the boat will start to move. After a while, the speed of the boat reaches its maximum value and the applied force is used up in overcoming the friction of the water against the hull.

When the battery switch (S_1) in the LR circuit of figure 2-10(A) is closed, the rate of the current increase is maximum in the inductive circuit. At this instant all the battery voltage is used in overcoming the emf of self-induction which is a maximum because the rate of change of current is maximum. Thus the battery voltage is equal to the drop across the inductor and the voltage across the resistor is zero. As time goes on more of the battery voltage appears across the resistor and less across the inductor. The rate of change of current is less and the induced emf is less. As the steady-state condition of the current is approached, the drop across the inductor approaches zero and all of the battery voltage is "dropped" across the resistance of the circuit.

Thus the voltages across the inductor and the resistor change in magnitude during the period of growth of current the same way the force applied to the boat divides itself between the effects of inertia and friction. In both examples, the force is developed first across the inertia/inductive effect and finally across the friction/resistive effect.

Figure 2-10(C) shows that when switch S_2 is closed (source voltage E_S removed from the circuit), the flux that has been established around the inductor (L) collapses through the windings. This induces a voltage e_L in the inductor that has a polarity opposite to E_S and is essentially equal to E_S in magnitude. The induced voltage causes decay current (i_d) to flow in resistor R in the same direction in which current was flowing originally (when S_1 was closed). A voltage (e_R) that is initially equal to source voltage (E_S) is developed across R. The voltage across the resistor (e_R) rapidly falls to zero as the voltage across the inductor (e_L) falls to zero due to the collapsing flux.

Just as the example of the boat was used to explain the growth of current in a circuit, it can also be used to explain the decay of current in a circuit. When the force applied to the boat is removed, the boat still continues to move through the water for a while, eventually coming to a stop. This is because energy was being stored in the inertia of the moving boat. After a period of time the friction of the water overcomes the inertia of the boat, and the boat stops moving. Just as inertia of the boat stored energy, the magnetic field of an inductor stores energy. Because of this, even when the power source is removed, the stored energy of the magnetic field of the inductor tends to keep current flowing in the circuit until the magnetic field collapse.

Q8.

    a. When voltage is first applied to a series LR circuit, how much opposition does the inductance have to the flow of current compared to that of the circuit resistance?

    b. In a series circuit containing a resistor (R_1) and an inductor (L_1), what voltage exists across R_1 when the counter emf is at its maximum value?

    c. What happens to the voltage across the resistance in an LR circuit during current buildup in the circuit, and during current decay in the circuit?

L/R Time Constant

The L/R TIME CONSTANT is a valuable tool for use in determining the time required for current in an inductor to reach a specific value. As shown in figure 2-11, one L/R time constant is the time required for the current in an inductor to increase to 63 percent (actually 63.2 percent) of the maximum current. Each time constant is equal to the time required for the current to increase by 63.2 percent of the difference in value between the current flowing in the inductor and the maximum current. Maximum current flows in the inductor after five L/R time constants are completed. The following example should clear up any confusion about time constants. Assume that maximum current in an LR circuit is 10 amperes. As you know, when the circuit is energized, it takes time for the current to go from zero to 10 amperes. When the first time constant is completed, the current in the circuit is equal to 63.2% of 10 amperes. Thus the amplitude of current at the end of 1 time constant is 6.32 amperes.

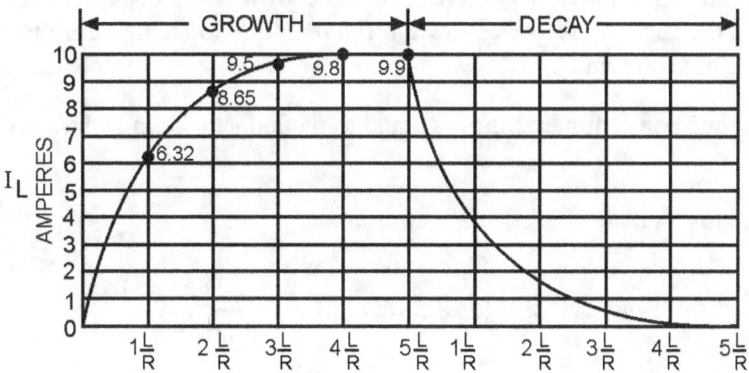

Figure 2-11.—L/R time constant.

During the second time constant, current again increases by 63.2% (.632) of the difference in value between the current flowing in the inductor and the maximum current. This difference is 10 amperes minus 6.32 amperes and equals 3.68 amperes; 63.2% of 3.68 amperes is 2.32 amperes. This increase in current during the second time constant is added to that of the first time constant. Thus, upon completion of the second time constant, the amount of current in the LR circuit is 6.32 amperes + 2.32 amperes = 8.64 amperes.

During the third constant, current again increases:

$$10 \text{ amperes} - 8.64 \text{ amperes} = 1.36 \text{ amperes}$$
$$1.36 \text{ amperes} \times .632 = 0.860 \text{ ampere}$$
$$8.64 \text{ amperes} + 0.860 \text{ ampere} = 9.50 \text{ amperes}$$

During the fourth time constant, current again increases:

$$10 \text{ amperes} - 9.50 \text{ amperes} = 0.5 \text{ ampere}$$
$$0.5 \text{ ampere} \times .632 = 0.316 \text{ ampere}$$
$$9.50 \text{ amperes} + 0.316 \text{ ampere} = 9.82 \text{ amperes}$$

During the fifth time constant, current increases as before:

$$10 \text{ amperes} - 9.82 \text{ amperes} = 0.18 \text{ ampere}$$
$$0.18 \text{ ampere} \times .632 = 0.114 \text{ ampere}$$
$$9.82 \text{ amperes} + .114 \text{ ampere} = 9.93 \text{ amperes}$$

Thus, the current at the end of the fifth time constant is almost equal to 10.0 amperes, the maximum current. For all practical purposes the slight difference in value can be ignored.

When an LR circuit is deenergized, the circuit current decreases (decays) to zero in five time constants at the same rate that it previously increased. If the growth and decay of current in an LR circuit are plotted on a graph, the curve appears as shown in figure 2-11. Notice that current increases and decays at the same rate in five time constants.

The value of the time constant in seconds is equal to the inductance in henrys divided by the circuit resistance in ohms.

The formula used to calculate one L/R time constant is:

$$\text{Time Constant (TC) in seconds} = \frac{L \text{ (in henrys)}}{R \text{ (in ohms)}}$$

Q9. *What is the formula for one L/R time constant?*

Q10.

a. *The maximum current applied to an inductor is 1.8 amperes. How much current flowed in the inductor 3 time constants after the circuit was first energized?*

b. *What is the minimum number of time constants required for the current in an LR circuit to increase to its maximum value?*

c. *A circuit containing only an inductor and a resistor has a maximum of 12 amperes of applied current flowing in it. After 5 L/R time constants the circuit is opened. How many time constants is required for the current to decay to 1.625 amperes?*

POWER LOSS IN AN INDUCTOR

Since an inductor (coil) consists of a number of turns of wire, and since all wire has some resistance, every inductor has a certain amount of resistance. Normally this resistance is small. It is usually neglected in solving various types of ac circuit problems because the reactance of the inductor (the opposition to alternating current, which will be discussed later) is so much greater than the resistance that the resistance has a negligible effect on the current.

However, since some inductors are designed to carry relatively large amounts of current, considerable power can be dissipated in the inductor even though the amount of resistance in the inductor is small. This power is wasted power and is called COPPER LOSS. The copper loss of an inductor can be calculated by multiplying the square of the current in the inductor by the resistance of the winding (I^2R).

In addition to copper loss, an iron-core coil (inductor) has two iron losses. These are called HYSTERESIS LOSS and EDDY-CURRENT LOSS. Hysteresis loss is due to power that is consumed in reversing the magnetic field of the inductor core each time the direction of current in the inductor changes.

Eddy-current loss is due to heating of the core by circulating currents that are induced in the iron core by the magnetic field around the turns of the coil. These currents are called eddy currents and circulate within the iron core only.

All these losses dissipate power in the form of heat. Since this power cannot be returned to the electrical circuit, it is lost power.

Q11. State three types of power loss in an inductor.

MUTUAL INDUCTANCE

Whenever two coils are located so that the flux from one coil links with the turns of the other coil, a change of flux in one coil causes an emf to be induced in the other coil. This allows the energy from one coil to be transferred or coupled to the other coil. The two coils are said to be coupled or linked by the property of MUTUAL INDUCTANCE (M). The amount of mutual inductance depends on the relative positions of the two coils. This is shown in figure 2-12. If the coils are separated a considerable distance, the amount of flux common to both coils is small and the mutual inductance is low. Conversely, if the coils are close together so that nearly all the flux of one coil links the turns of the other, the mutual inductance is high. The mutual inductance can be increased greatly by mounting the coils on a common iron core.

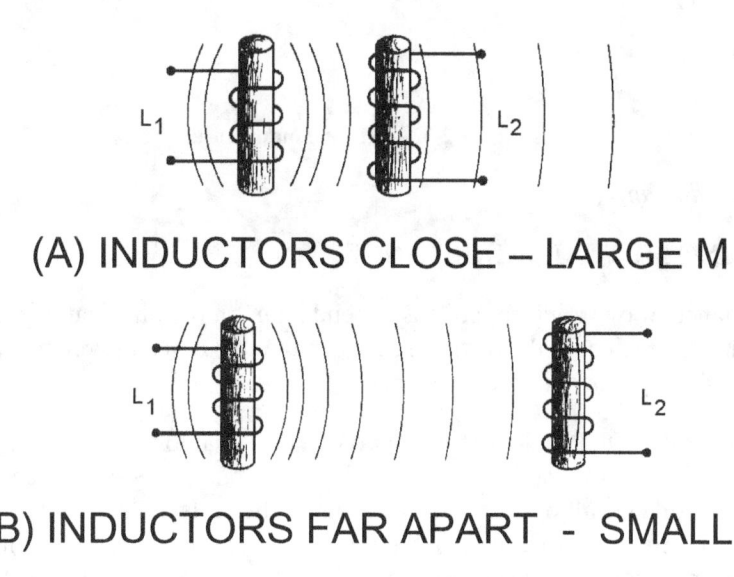

(A) INDUCTORS CLOSE – LARGE M

(B) INDUCTORS FAR APART - SMALL M

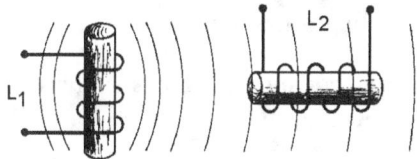

(C) INDUCTOR AXES PERPENDICULAR – NO M

Figure 2-12.—The effect of position of coils on mutual inductance (M).

Two coils are placed close together as shown in figure 2-13. Coil 1 is connected to a battery through switch S, and coil 2 is connected to an ammeter (A). When switch S is closed as in figure 2-13(A), the current that flows in coil 1 sets up a magnetic field that links with coil 2, causing an induced voltage in coil 2 and a momentary deflection of the ammeter. When the current in coil 1 reaches a steady value, the ammeter returns to zero. If switch S is now opened as in figure 2-13(B), the ammeter (A) deflects momentarily in the opposite direction, indicating a momentary flow of current in the opposite direction in coil 2. This current in coil 2 is produced by the collapsing magnetic field of coil 1.

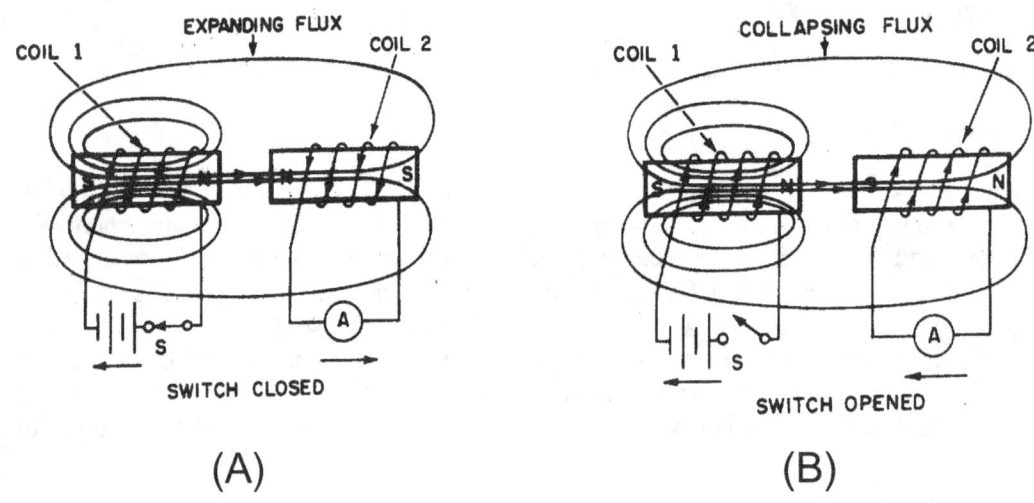

Figure 2-13.—Mutual inductance.

Q12. Define mutual inductance.

FACTORS AFFECTING MUTUAL INDUCTANCE

The mutual inductance of two adjacent coils is dependent upon the physical dimensions of the two coils, the number of turns in each coil, the distance between the two coils, the relative positions of the axes of the two coils, and the permeability of the cores.

The COEFFICIENT OF COUPLING between two coils is equal to the ratio of the flux cutting one coil to the flux originated in the other coil. If the two coils are so positioned with respect to each other so that all of the flux of one coil cuts all of the turns of the other, the coils are said to have a unity coefficient of coupling. It is never exactly equal to unity (1), but it approaches this value in certain types of coupling devices. If all of the flux produced by one coil cuts only half the turns of the other coil, the coefficient of coupling is 0.5. The coefficient of coupling is designated by the letter K.

The mutual inductance between two coils, L_1 and L_2, is expressed in terms of the inductance of each coil and the coefficient of coupling K. As a formula:

$$M = K\sqrt{L_1 L_2}$$

where: M = Mutual inductance in henrys
K = Coefficient of coupling
L_1, L_2 = Inductance of coil in henrys

Example problem:

One 10-H coil and one 20-H coil are connected in series and are physically close enough to each other so that their coefficient of coupling is 0.5. What is the mutual inductance between the coils?

Use the formula:
$$M = K\sqrt{L_1 L_2}$$
$$M = 0.5\sqrt{(10H)(20H)}$$
$$M = 0.5\sqrt{200H}$$
$$M = 0.5 \times 14.14H$$
$$M = 7.07H$$

Q13. When are two circuits said to be coupled?

Q14. What is meant by the coefficient of coupling?

SERIES INDUCTORS WITHOUT MAGNETIC COUPLING

When inductors are well shielded or are located far enough apart from one another, the effect of mutual inductance is negligible. If there is no mutual inductance (magnetic coupling) and the inductors are connected in series, the total inductance is equal to the sum of the individual inductances. As a formula:

$$L_T = L_1 + L_2 + L_3 + \ldots L_n$$

where L_T is the total inductance; L_1, L_2, L_3 are the inductances of L_1, L_2, L_3; and L_n means that any number (n) of inductors may be used. The inductances of inductors in series are added together like the resistances of resistors in series.

SERIES INDUCTORS WITH MAGNETIC COUPLING

When two inductors in series are so arranged that the field of one links the other, the combined inductance is determined as follows:

$$L_T = L_1 + L_2 \pm 2M$$

where: L_T = The total inductance
L_1, L_2 = The inductances of L_1, L_2
M = The mutual inductance between the two inductors

The plus sign is used with M when the magnetic fields of the two inductors are aiding each other, as shown in figure 2-14. The minus sign is used with M when the magnetic field of the two inductors oppose each other, as shown in figure 2-15. The factor 2M accounts for the influence of L_1 on L_2 and L_2 on L_1.

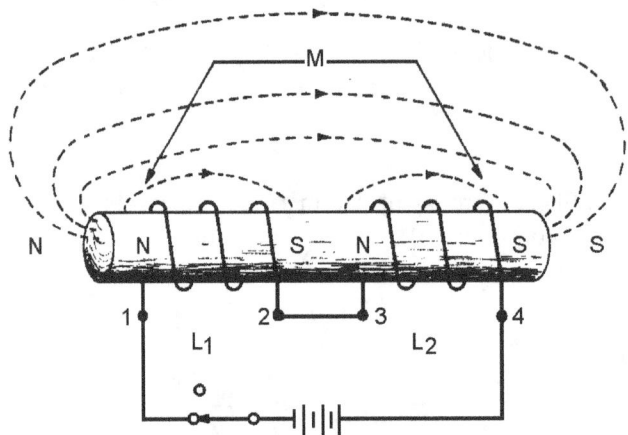

Figure 2-14.—Series inductors with aiding fields.

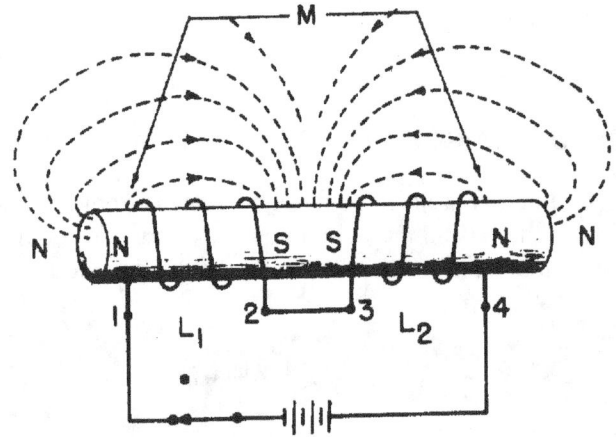

Figure 2-15.—Series inductors with opposing fields.

Example problem:

A 10-H coil is connected in series with a 5-H coil so the fields aid each other. Their mutual inductance is 7 H. What is the combined inductance of the coils?

Use the formula: $L_T = L_1 + L_2 + 2M$

$L_T = 10\text{ H} + 5\text{ H} + 2(7\text{ H})$

$L_T = 29\text{ H}$

Q15. *Two series-connected 7-H inductors are adjacent to each other; their coefficient of coupling is 0.64. What is the value of M?*

Q16. *A circuit contains two series inductors aligned in such a way that their magnetic fields oppose each other. What formula should you use to compute total inductance in this circuit?*

Q17. *The magnetic fields of two coils are aiding each other. The inductance of the coils are 3 H and 5 H, respectively. The coils' mutual inductance is 5 H. What is their combined inductance?*

PARALLEL INDUCTORS WITHOUT COUPLING

The total inductance (L_T) of inductors in parallel is calculated in the same manner that the total resistance of resistors in parallel is calculated, provided the coefficient of coupling between the coils is zero. Expressed mathematically:

$$\frac{1}{L_T} = \frac{1}{L_1} + \frac{1}{L_2} + \frac{1}{L_3} \ldots + \frac{1}{L_N}$$

SUMMARY

The important points of this chapter are summarized below. Study this information before continuing, as this information will lay the foundation for later chapters.

INDUCTANCE—The characteristic of an electrical circuit that opposes a change in current. The reaction (opposition) is caused by the creation or destruction of a magnetic field. When current starts to flow, magnetic lines of force are created. These lines of force cut the conductor inducing a counter emf in a <u>direction that opposes current</u>.

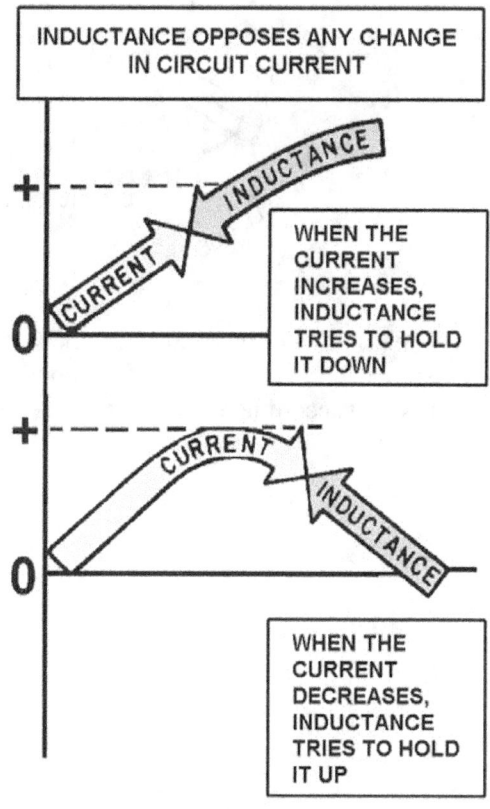

SELF-INDUCTANCE—The process by which a circuit induces an emf into itself by its own moving magnetic field. All electrical circuits possess self-inductance. This opposition (inductance), however, only takes place when there is a change in current. Inductance does NOT oppose current, only a CHANGE in current. The property of inductance can be increased by forming the conductor into a loop. In a loop, the magnetic lines of force affect more of the conductor at one time. This increases the self-induced emf.

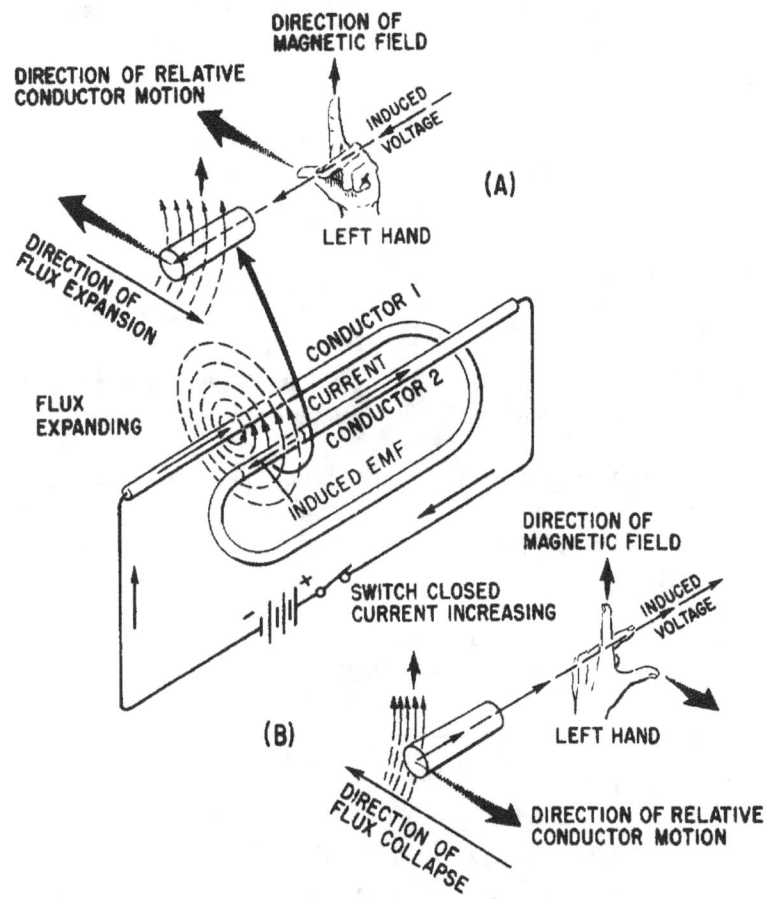

INDUCTANCE OF A COIL—The property of inductance can be further increased if the conductor is formed into a coil. Because a coil contains more loops, more of the conductor can be affected by the magnetic field. Inductors (coils) are classified according to core type. The core material is normally either air or soft iron.

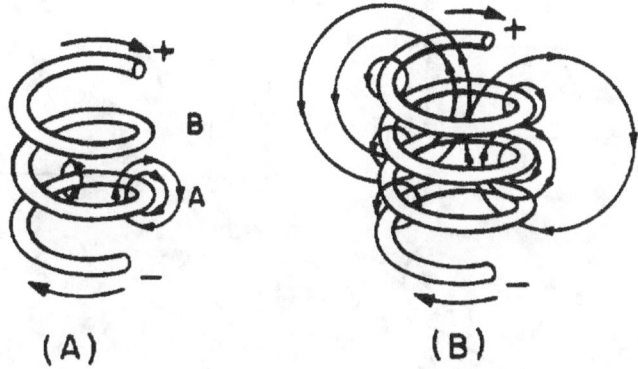

FACTORS AFFECTING COIL INDUCTANCE—The inductance of a coil is entirely dependent upon its physical construction. Some of the factors affecting the inductance are:

- The number of turns in the coil. Increasing the number of turns will increase the inductance.

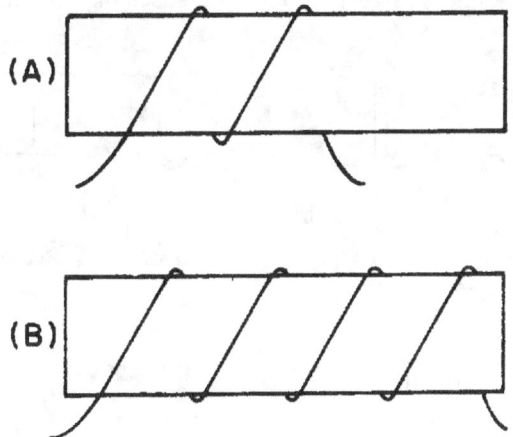

- The coil diameter. The inductance increases directly as the cross-sectional area of the coil increases.

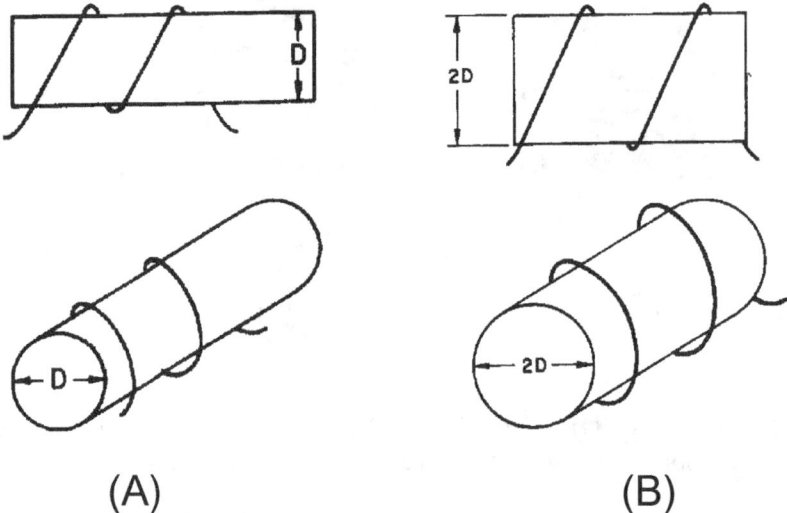

- The length of the coil. When the length of the coil is increased while keeping the number of turns the same, the turn-spacing is increased. This decreases the inductance of the coil.

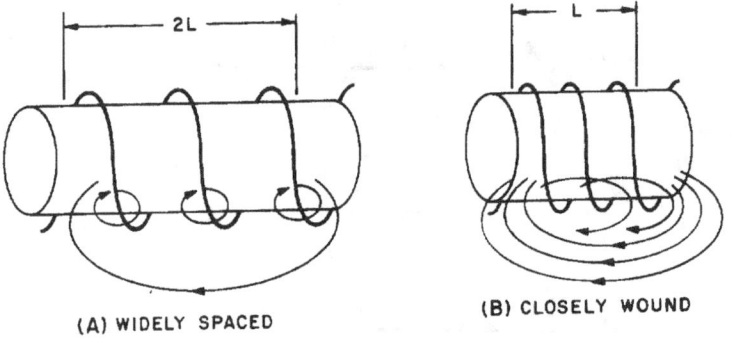

- The type of core material. Increasing the permeability of the core results in increasing the inductance of the coil.

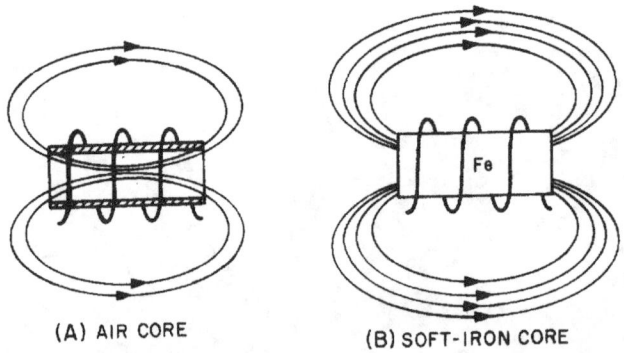

- Winding the coil in layers. The more layers used to form a coil, the greater effect the magnetic field has on the conductor. By layering a coil, you can increase the inductance.

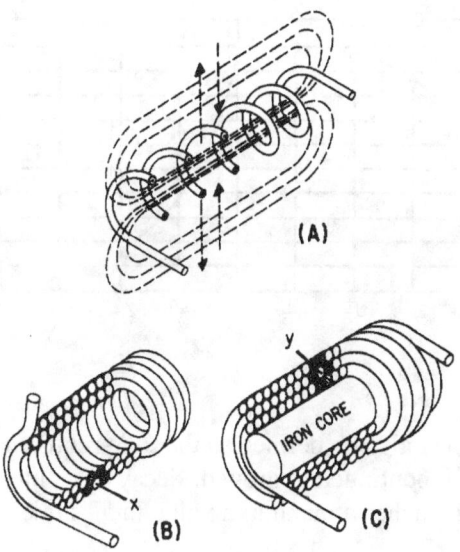

UNIT OF INDUCTANCE—Inductance (L) is measured in henrys (H). An inductor has an inductance of one henry (H) if an emf of one volt is induced in the inductor when the current through the inductor is changing at the rate of 1 ampere per second. Common units of inductance are henry (H), millihenry (mH) and the microhenry (μH).

GROWTH AND DECAY OF CURRENT IN AN LR CIRCUIT—The required for the current in an inductor to increase to 63.2 percent of the maximum current or to decrease to 36.8 percent of the maximum current is known as the time constant. The letter symbol for an LR time constant is L/R. As a formula:

$$\frac{L}{R}.$$

As a formula:

$$t \text{ (in seconds)} = \frac{L \text{ (in henrys)}}{R \text{ (in ohms)}}$$

or

$$t \text{ (in microseconds)} = \frac{L \text{ (in microhenrys)}}{R \text{ (in ohms)}}$$

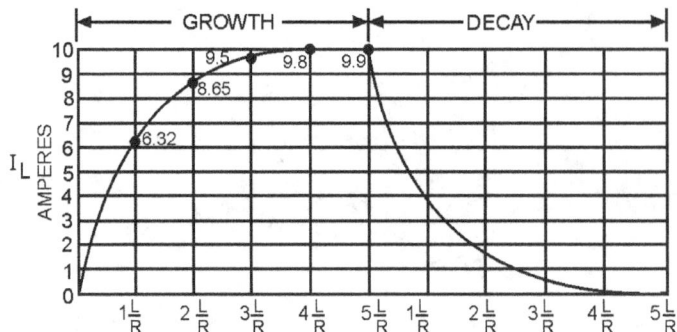

The time constant of an LR circuit may also be defined as the time required for the current in the inductor to grow or decay to its final value if it continued to grow or decay at its initial rate. For all practical purposes, the current in the inductor reaches a maximum value in 5 "Time Constants" and decreases to zero in 5 "Time Constants".

POWER LOSSES IN AN INDUCTOR—Since an inductor (coil) contains a number turns of wire, and all wire has some resistance, the inductor has a certain amount of resistance. This resistance is usually very small and has a negligible effect on current. However, there are power losses in an inductor. The main power losses in an inductor are copper loss, hysteresis loss, and eddy-current loss. Copper loss can be calculated by multiplying the square of the current by the resistance of the wire in the coil (I^2R). Hysteresis loss is due to power that is consumed in reversing the magnetic field of the core each time the current direction changes. Eddy-current loss is due to core heating caused by circulating currents induced in an iron core by the magnetic field of the coil.

MUTUAL INDUCTANCE—When two coils are located so that the flux from one coil cuts the turns of the other coil, the coils have mutual inductance. The amount of mutual inductance depends upon several factors: the relative position of the axes of the two coils; the permeability of the cores; the physical dimensions of the two coils; the number of turns in each coil, and the distance between the coils. The coefficient of coupling K specifies the amount of coupling between the coils. If all of the flux from one coil cuts all of the turns of the other coil, the coefficient of coupling K is 1 or unity. If none of the flux from one coil cuts the turns of the other coil, the coefficient of coupling is zero. The mutual inductance between two coils (L_1 and L_2) may be expressed mathematically as:

$$M = K\sqrt{L_1 L_2}$$

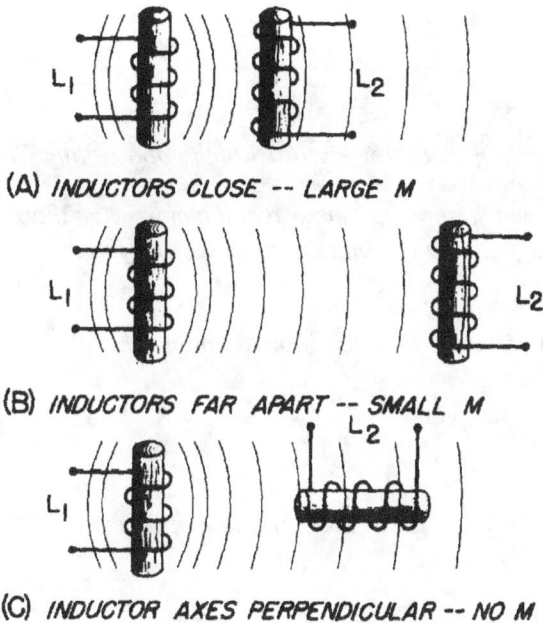

(A) INDUCTORS CLOSE -- LARGE M

(B) INDUCTORS FAR APART -- SMALL M

(C) INDUCTOR AXES PERPENDICULAR -- NO M

COMPUTING THE INDUCTANCE OF A CIRCUIT—When the total inductance of a circuit is computed, the individual inductive values are treated the same as resistance values. The inductances of inductors in series are added like the resistances of resistors in series. That is,

$$L_T = L_1 + L_2 + L_3 \ldots + L_n$$

The inductances of inductors in parallel are combined mathematically like the resistances of resistors in parallel. That is,

$$L_T = \frac{1}{\frac{1}{L_1} + \frac{1}{L_2} + \frac{1}{L_3} \ldots \frac{1}{L_n}}$$

Both of the above formulas are accurate, providing there is no mutual inductance between the inductors.

ANSWERS TO QUESTIONS Q1. THROUGH Q17.

A1. The henry, H.

A2. Magnetic field.

A3. Inductance is the property of a coil (or circuit) which opposes any CHANGE in current.

A4. Induced emf is the emf which appears across a conductor when there is relative motion between the conductor and a magnetic field; counter emf is the emf induced in a conductor that opposes the applied voltage.

A5. The induced emf in any circuit is in a direction to oppose the effect that produced it.

A6.
    a. No effect.
    b. Inductance opposes any change in the amplitude of current.

A7.
    a.
        1. The numbers of turns in a coil.
        2. The type of material used in the core.
        3. The diameter of the coil.
        4. The coil length.
        5. The number of layers of windings in the coil.
    b. Increases inductance.
    c. Increases inductance.
    d. Decreases inductance.
    e. Increases inductance.
    f. Increases inductance.

A8.
    a. Inductance causes a very large opposition to the flow of current when voltage is first applied to an LR circuit; resistance causes comparatively little opposition to current at that time.
    b. Zero.
    c. During current buildup, the voltage across the resistor gradually increases to the same voltage as the source voltage; and during current decay the voltage across the resistor gradually drops to zero

A9.

$$t = \frac{L}{R}$$

A10.

    a.   1.71 amperes.

    b.   5 time constants.

    c.   2 time constants.

A11.   Copper loss; hysteresis loss; eddy-current loss.

A12.   Mutual inductance is the property existing between two coils so positioned that flux from one coil cuts the windings of the other coil.

A13.   When they are arranged so that energy from one circuit is transferred to the other circuit.

A14.   The ratio of the fines of force produced by one coil to the lines of force that link another coil. It is never greater than one.

A15.

$$4.48\ H\ (\text{because } M = K\sqrt{7H \times 7H} = 0.64 \times 7H = 4.48H)$$

A16.

$$L_T = L_1 + L_2 - 2M$$

A17.

$$L_T = 18\ H\ (\text{because } L_T = L_1 + L_2 + 2M$$
$$L_T = 3H + 5H + 2(5H) = 18H)$$

CHAPTER 3
CAPACITANCE

LEARNING OBJECTIVES

Upon completion of this chapter you will be able to:

1. Define the terms "capacitor" and "capacitance."

2. State four characteristics of electrostatic lines of force.

3. State the effect that an electrostatic field has on a charged particle.

4. State the basic parts of a capacitor.

5. Define the term "farad".

6. State the mathematical relationship between a farad, a microfarad, and a picofarad.

7. State three factors that affect the value of capacitance.

8. Given the dielectric constant and the area of and the distance between the plates of a capacitor, solve for capacitance.

9. State two types of power losses associated with capacitors.

10. Define the term "working voltage" of a capacitor, and compute the working voltage of a capacitor.

11. State what happens to the electrons in a capacitor when the capacitor is charging and when it is discharging.

12. State the relationship between voltage and time in an RC circuit when the circuit is charging and discharging.

13. State the relationship between the voltage drop across a resistor and the source voltage in an RC circuit.

14. Given the component values of an RC circuit, compute the RC time constant.

15. Use the universal time constant chart to determine the value of an unknown capacitor in an RC circuit.

16. Calculate the value of total capacitance in a circuit containing capacitors of known value in series.

17. Calculate the value of total capacitance in a circuit containing capacitors of known value in parallel.

18. State the difference between different types of capacitors.

19. Determine the electrical values of capacitors using the color code.

CAPACITANCE

In the previous chapter you learned that inductance is the property of a coil that causes electrical energy to be stored in a magnetic field about the coil. The energy is stored in such a way as to <u>oppose any change in current</u>. CAPACITANCE is similar to inductance because it also causes a storage of energy. A CAPACITOR is a device that stores electrical energy in an <u>ELECTROSTATIC FIELD</u>. The energy is stored in such a way as to <u>oppose any change in voltage</u>. Just how capacitance opposes a change in voltage is explained later in this chapter. However, it is first necessary to explain the principles of an electrostatic field as it is applied to capacitance.

Q1. Define the terms "capacitor" and "capacitance."

THE ELECTROSTATIC FIELD

You previously learned that opposite electrical charges attract each other while like electrical charges repel each other. The reason for this is the existence of an electrostatic field. Any charged particle is surrounded by invisible lines of force, called electrostatic lines of force. These lines of force have some interesting characteristics:

- They are polarized from positive to negative.

- They radiate from a charged particle in straight lines and do not form closed loops.

- They have the ability to pass through any known material.

- They have the ability to distort the orbits of tightly bound electrons.

Examine figure 3-1. This figure represents two unlike charges surrounded by their electrostatic field. Because an electrostatic field is polarized positive to negative, arrows are shown radiating <u>away</u> from the positive charge and <u>toward</u> the negative charge. Stated another way, the field from the positive charge is pushing, while the field from the negative charge is pulling. The effect of the field is to push and pull the unlike charges together.

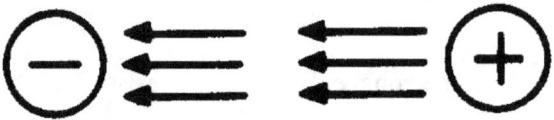

Figure 3-1.—Electrostatic field attracts two unlike charged particles.

In figure 3-2, two like charges are shown with their surrounding electrostatic field. The effect of the electrostatic field is to push the charges apart.

Figure 3-2.—Electrostatic field repels two like charged particles.

If two unlike charges are placed on opposite sides of an atom whose outermost electrons cannot escape their orbits, the orbits of the electrons are distorted as shown in figure 3-3. Figure 3-3(A) shows the normal orbit. Part (B) of the figure shows the same orbit in the presence of charged particles. Since the electron is a negative charge, the positive charge attracts the electrons, pulling the electrons closer to the positive charge. The negative charge repels the electrons, pushing them further from the negative charge. It is this ability of an electrostatic field to attract and to repel charges that allows the capacitor to store energy.

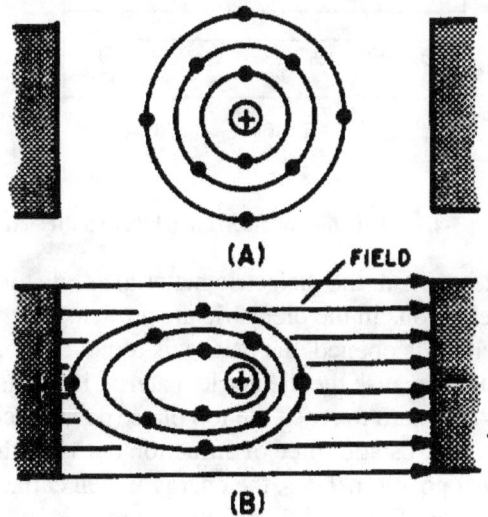

Figure 3-3.—Distortion of electron orbital paths due to electrostatic force.

Q2. *State four characteristics of electrostatic lines of force.*

Q3. *An electron moves into the electrostatic field between a positive charge and a negative charge. Toward which charge will the electron move?*

THE SIMPLE CAPACITOR

A simple capacitor consists of two metal plates separated by an insulating material called a <u>dielectric</u>, as illustrated in figure 3-4. Note that one plate is connected to the positive terminal of a battery; the other plate is connected through a closed switch (S1) to the negative terminal of the battery. Remember, an insulator is a material whose electrons cannot easily escape their orbits. Due to the battery voltage, plate A is charged positively and plate B is charged negatively. (How this happens is explained later in this chapter.) Thus an electrostatic field is set up between the positive and negative plates. The electrons on the negative plate (plate B) are attracted to the positive charges on the positive plate (plate A).

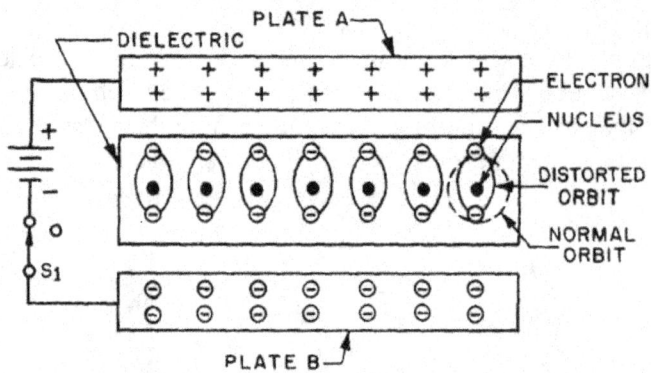

Figure 3-4.—Distortion of electron orbits in a dielectric.

Notice that the orbits of the electrons in the dielectric material are distorted by the electrostatic field. The distortion occurs because the electrons in the dielectric are attracted to the top plate while being repelled from the bottom plate. When switch S1 is opened, the battery is removed from the circuit and the charge is retained by the capacitor. This occurs because the dielectric material is an insulator, and the electrons in the bottom plate (negative charge) have no path to reach the top plate (positive charge). The distorted orbits of the atoms of the dielectric plus the electrostatic force of attraction between the two plates hold the positive and negative charges in their original position. Thus, the energy which came from the battery is now stored in the electrostatic field of the capacitor. Two slightly different symbols for representing a capacitor are shown in figure 3-5. Notice that each symbol is composed of two plates separated by a space that represents the dielectric. The curved plate in (B) of the figure indicates the plate should be connected to a negative polarity.

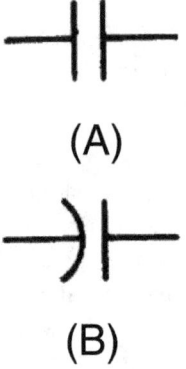

Figure 3-5.—Circuit symbols for capacitors.

Q4. What are the basic parts of a capacitor?

THE FARAD

Capacitance is measured in units called FARADS. A one-farad capacitor stores one coulomb (a unit of charge (Q) equal to 6.28×10^{18} electrons) of charge when a potential of 1 volt is applied across the terminals of the capacitor. This can be expressed by the formula:

$$C(\text{farads}) = \frac{Q(\text{coulombs})}{E(\text{volts})}$$

The farad is a very large unit of measurement of capacitance. For convenience, the microfarad (abbreviated μF) or the picofarad (abbreviated μF) is used. One (1.0) microfarad is equal to 0.000001 farad or 1×10^{-6} farad, and 1.0 picofarad is equal to 0.000000000001 farad or 1.0×10^{-12} farad. Capacitance is a physical property of the capacitor and does not depend on circuit characteristics of voltage, current, and resistance. A given capacitor always has the same value of capacitance (farads) in one circuit as in any other circuit in which it is connected.

Q5. Define the term "farad."

Q6. What is the mathematical relationship between a farad, a microfarad, and a picofarad.

FACTORS AFFECTING THE VALUE OF CAPACITANCE

The value of capacitance of a capacitor depends on three factors:

- The area of the plates.

- The distance between the plates.

- The dielectric constant of the material between the plates.

PLATE AREA affects the value of capacitance in the same manner that the size of a container affects the amount of water that can be held by the container. A capacitor with the large plate area can store more charges than a capacitor with a small plate area. Simply stated, "the larger the plate area, the larger the capacitance".

The second factor affecting capacitance is the DISTANCE BETWEEN THE PLATES. Electrostatic lines of force are strongest when the charged particles that create them are close together. When the charged particles are moved further apart, the lines of force weaken, and the ability to store a charge decreases.

The third factor affecting capacitance is the DIELECTRIC CONSTANT of the insulating material between the plates of a capacitor. The various insulating materials used as the dielectric in a capacitor differ in their ability to respond to (pass) electrostatic lines of force. A dielectric material, or insulator, is rated as to its ability to respond to electrostatic lines of force in terms of a figure called the DIELECTRIC CONSTANT. A dielectric material with a high dielectric constant is a better insulator than a dielectric material with a low dielectric constant. Dielectric constants for some common materials are given in the following list:

Material	Constant
Vacuum	1.0000
Air	1.0006
Paraffin paper	3.5
Glass	5 to 10
Mica	3 to 6
Rubber	2.5 to 35
Wood	2.5 to 8
Glycerine (15°C)	56
Petroleum	2
Pure water	81

Notice the dielectric constant for a vacuum. Since a vacuum is the standard of reference, it is assigned a constant of one. The dielectric constants of all materials are compared to that of a vacuum. Since the dielectric constant of air has been determined to be approximately the same as that of a vacuum, the dielectric constant of AIR is also considered to be equal to one.

The formula used to compute the value of capacitance is:

$$C = 0.2249 \left(\frac{KA}{d}\right)$$

Where C = capacitance in picofarads

A = area of one plate, in square inches

d = distance between the plates, in inches

K = dielectric constant of the insulating material

0.2249 = a constant resulting from conversion from Metric to English units.

Example: Find the capacitance of a parallel plate capacitor with paraffin paper as the dielectric.

Given: K = 3.5
         d = 0.05 inch
         A = 12 square inches

Solution: $C = 0.2249\left(\dfrac{KA}{d}\right)$

$C = 0.2249\left(\dfrac{3.5 \times 12}{0.05}\right)$

$C = 189 \text{ picofarads}$

By examining the above formula you can see that capacitance varies directly as the dielectric constant and the area of the capacitor plates, and inversely as the distance between the plates.

Q7. *State three factors that affect the capacitance of a capacitor.*

Q8. *A parallel plate capacitor has the following values: K = 81, d = .025 inches, A = 6 square inches. What is the capacitance of the capacitor?*

VOLTAGE RATING OF CAPACITORS

In selecting or substituting a capacitor for use, consideration must be given to (1) the value of capacitance desired and (2) the amount of voltage to be applied across the capacitor. If the voltage applied across the capacitor is too great, the dielectric will break down and arcing will occur between the capacitor plates. When this happens the capacitor becomes a short-circuit and the flow of direct current through it can cause damage to other electronic parts. Each capacitor has a voltage rating (a working voltage) that should not be exceeded.

The working voltage of the capacitor is the maximum voltage that can be steadily applied without danger of breaking down the dielectric. The working voltage depends on the type of material used as the dielectric and on the thickness of the dialectic. (A high-voltage capacitor that has a thick dielectric must have a relatively large plate area in order to have the same capacitance as a similar low-voltage capacitor having a thin dielectric.) The working voltage also depends on the applied frequency because the losses, and the resultant heating effect, <u>increase as the frequency increases</u>.

A capacitor with a voltage rating of 500 volts dc cannot be safely subjected to an alternating voltage or a pulsating direct voltage having an effective value of 500 volts. Since an alternating voltage of 500 volts (rms) has a peak value of 707 volts, a capacitor to which it is applied should have a working voltage of at least 750 volts. In practice, a capacitor should be selected so that its working voltage is at least 50 percent greater than the highest effective voltage to be applied to it.

CAPACITOR LOSSES

Power loss in a capacitor may be attributed to dielectric hysteresis and dielectric leakage. Dielectric hysteresis may be defined as an effect in a dielectric material similar to the hysteresis found in a magnetic material. It is the result of changes in orientation of electron orbits in the dielectric because of the rapid reversals of the polarity of the line voltage. The amount of power loss due to dielectric hysteresis depends upon the type of dielectric used. A vacuum dielectric has the smallest power loss.

Dielectric leakage occurs in a capacitor as the result of LEAKAGE CURRENT through the dielectric. Normally it is assumed that the dielectric will effectively prevent the flow of current through the capacitor. Although the resistance of the dielectric is extremely high, a minute amount of current does flow. Ordinarily this current is so small that for all practical purposes it is ignored. However, if the leakage through the dielectric is abnormally high, there will be a rapid loss of charge and an overheating of the capacitor.

The power loss of a capacitor is determined by loss in the dielectric. If the loss is negligible and the capacitor returns the total charge to the circuit, it is considered to be a perfect capacitor with a power loss of zero.

Q9. Name two types of power losses associated with a capacitor.

Q10.

    a. Define the term "working voltage" of a capacitor.

    b. What should be the working voltage of a capacitor in a circuit that is operating at 600 volts?

CHARGING AND DISCHARGING A CAPACITOR

CHARGING

In order to better understand the action of a capacitor in conjunction with other components, the charge and discharge actions of a purely capacitive circuit are analyzed first. For ease of explanation the capacitor and voltage source shown in figure 3-6 are assumed to be perfect (no internal resistance), although this is impossible in practice.

In figure 3-6(A), an uncharged capacitor is shown connected to a four-position switch. With the switch in position 1 the circuit is open and no voltage is applied to the capacitor. Initially each plate of the capacitor is a neutral body and until a difference of potential is impressed across the capacitor, no electrostatic field can exist between the plates.

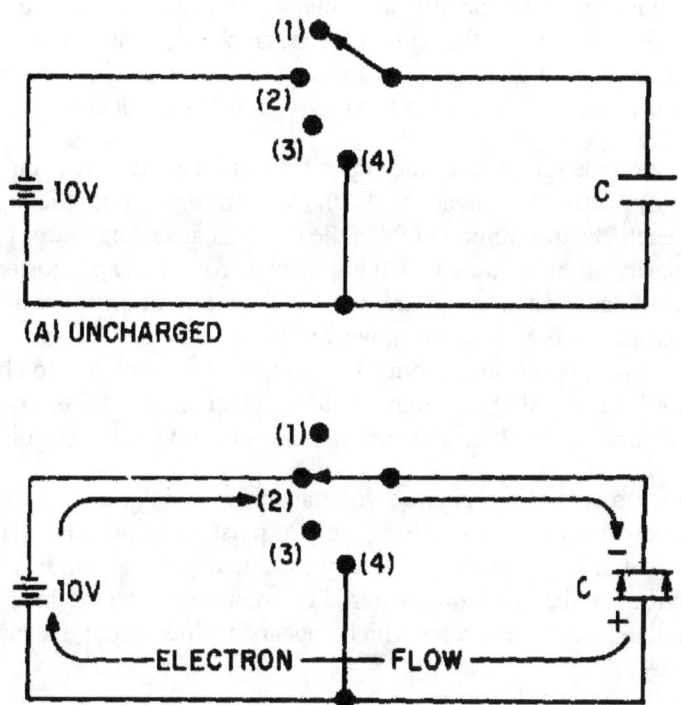

Figure 3-6.—Charging a capacitor.

To CHARGE the capacitor, the switch must be thrown to position 2, which places the capacitor across the terminals of the battery. Under the assumed perfect conditions, the capacitor would reach full charge instantaneously. However, the charging action is spread out over a period of time in the following discussion so that a step-by-step analysis can be made.

At the instant the switch is thrown to position 2 (fig. 3-6(B)), a displacement of electrons occurs simultaneously in all parts of the circuit. This electron displacement is directed away from the negative terminal and toward the positive terminal of the source (the battery). A brief surge of current will flow as the capacitor charges.

If it were possible to analyze the motion of the individual electrons in this surge of charging current, the following action would be observed. See figure 3-7.

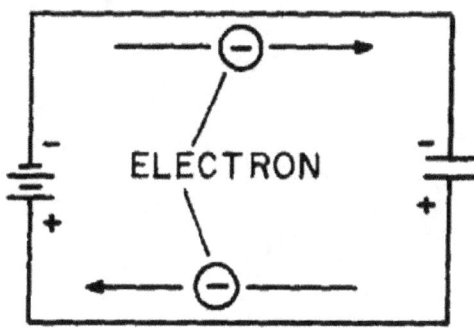

Figure 3-7.—Electron motion during charge.

3-9

At the instant the switch is closed, the positive terminal of the battery extracts an electron from the bottom conductor. The negative terminal of the battery forces an electron into the top conductor. At this same instant an electron is forced into the top plate of the capacitor and another is pulled from the bottom plate. Thus, in every part of the circuit a clockwise DISPLACEMENT of electrons occurs simultaneously.

As electrons accumulate on the top plate of the capacitor and others depart from the bottom plate, a difference of potential develops across the capacitor. Each electron forced onto the top plate makes that plate more negative, while each electron removed from the bottom causes the bottom plate to become more positive. Notice that the polarity of the voltage which builds up across the capacitor is such as to oppose the source voltage. The source voltage (emf) forces current around the circuit of figure 3-7 in a clockwise direction. The emf developed across the capacitor, however, has a tendency to force the current in a counterclockwise direction, opposing the source emf. As the capacitor continues to charge, the voltage across the capacitor rises until it is equal to the source voltage. Once the capacitor voltage equals the source voltage, the two voltages balance one another and current ceases to flow in the circuit.

In studying the charging process of a capacitor, you must be aware that NO current flows THROUGH the capacitor. The material between the plates of the capacitor must be an insulator. However, to an observer stationed at the source or along one of the circuit conductors, the action has all the appearances of a true flow of current, even though the insulating material between the plates of the capacitor prevents the current from having a complete path. The current which appears to flow through a capacitor is called DISPLACEMENT CURRENT.

When a capacitor is fully charged and the source voltage is equaled by the counter electromotive force (cemf) across the capacitor, the electrostatic field between the plates of the capacitor is maximum. (Look again at figure 3-4.) Since the electrostatic field is maximum the energy stored in the dielectric is also maximum.

If the switch is now opened as shown in figure 3-8(A), the electrons on the upper plate are isolated. The electrons on the top plate are attracted to the charged bottom plate. Because the dielectric is an insulator, the electrons can not cross the dielectric to the bottom plate. The charges on both plates will be effectively trapped by the electrostatic field and the capacitor will remain charged indefinitely. You should note at this point that the insulating dielectric material in a practical capacitor is not perfect and small leakage current will flow through the dielectric. This current will eventually dissipate the charge. However, a high quality capacitor may hold its charge for a month or more.

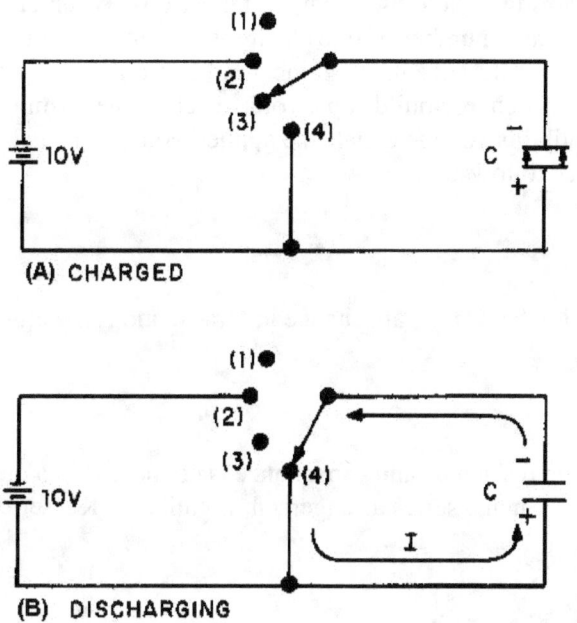

Figure 3-8.—Discharging a capacitor.

To review briefly, when a capacitor is connected across a voltage source, a surge of charging current flows. This charging current develops a cemf across the capacitor which opposes the applied voltage. When the capacitor is fully charged, the cemf is equal to the applied voltage and charging current ceases. At full charge, the electrostatic field between the plates is at maximum intensity and the energy stored in the dielectric is maximum. If the charged capacitor is disconnected from the source, the charge will be retained for some period of time. The length of time the charge is retained depends on the amount of leakage current present. Since electrical energy is stored in the capacitor, a charged capacitor can act as a source emf.

DISCHARGING

To DISCHARGE a capacitor, the charges on the two plates must be neutralized. This is accomplished by providing a conducting path between the two plates as shown in figure 3-8(B). With the switch in position (4) the excess electrons on the negative plate can flow to the positive plate and neutralize its charge. When the capacitor is discharged, the distorted orbits of the electrons in the dielectric return to their normal positions and the stored energy is returned to the circuit. It is important for you to note that a capacitor does not consume power. The energy the capacitor draws from the source is recovered when the capacitor is discharged.

Q11. State what happens to the electrons in a capacitor circuit when (a) the capacitor is charging and (b) the capacitor is discharging.

CHARGE AND DISCHARGE OF AN RC SERIES CIRCUIT

Ohm's law states that the voltage across a resistance is equal to the current through the resistance times the value of the resistance. This means that a voltage is developed across a resistance ONLY WHEN CURRENT FLOWS through the resistance.

A capacitor is capable of storing or holding a charge of electrons. When uncharged, both plates of the capacitor contain essentially the same number of free electrons. When charged, one plate contains more free electrons than the other plate. The difference in the number of electrons is a measure of the <u>charge</u> on the capacitor. The accumulation of this charge builds up a voltage across the terminals of the capacitor, and the charge continues to increase until this voltage equals the applied voltage. The <u>charge</u> in a capacitor is related to the capacitance and voltage as follows:

$$Q = CE,$$

in which Q is the charge in coulombs, C the capacitance in farads, and E the emf across the capacitor in volts.

CHARGE CYCLE

A voltage divider containing resistance and capacitance is connected in a circuit by means of a switch, as shown at the top of figure 3-9. Such a series arrangement is called an RC series circuit.

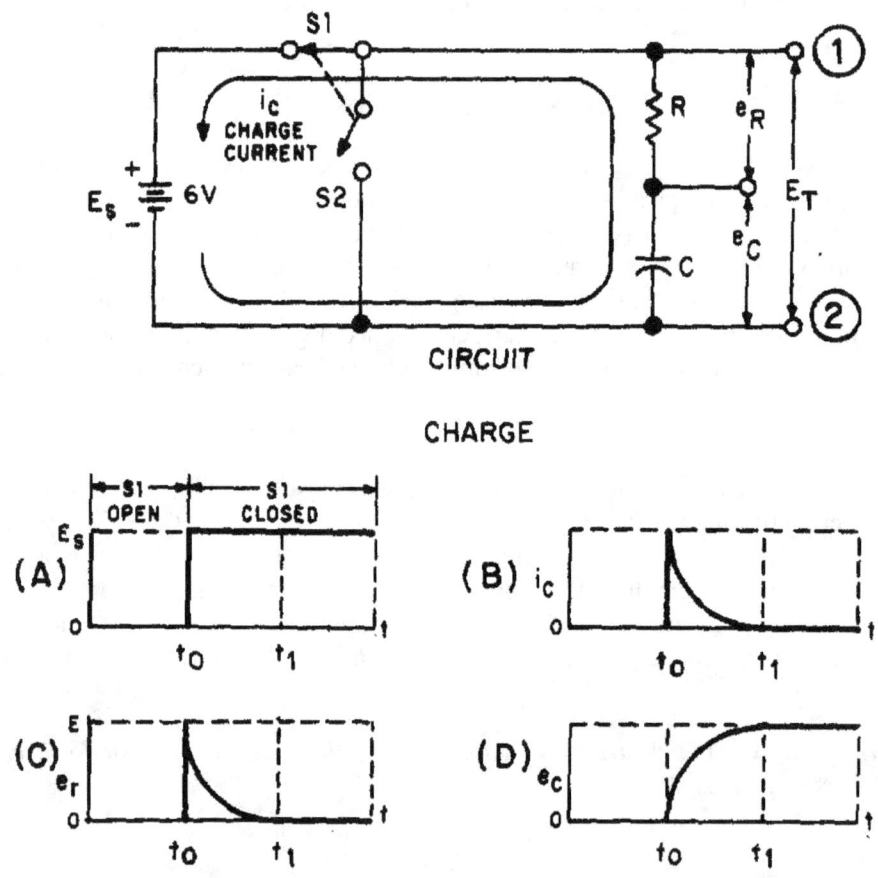

Figure 3-9.—Charge of an RC series circuit.

In explaining the charge and discharge cycles of an RC series circuit, the time interval from time t_0 (time zero, when the switch is first closed) to time t_1 (time one, when the capacitor reaches full charge or discharge potential) will be used. (Note that switches S1 and S2 move at the same time and can never both be closed at the same time.)

When switch S1 of the circuit in figure 3-9 is closed at t_0, the source voltage (E_S) is instantly felt across the entire circuit. Graph (A) of the figure shows an instantaneous rise at time t_0 from zero to source voltage (E_S = 6 volts). The total voltage can be measured across the circuit between points 1 and 2. Now look at graph (B) which represents the charging current in the capacitor (i_c). At time t_0, charging current is MAXIMUM. As time elapses toward time t_1, there is a continuous decrease in current flowing into the capacitor. The decreasing flow is caused by the voltage buildup across the capacitor. At time t_1, current flowing in the capacitor stops. At this time, the capacitor has reached full charge and has stored maximum energy in its electrostatic field. Graph (C) represents the voltage drop (e) across the resistor (R). The value of e_r is determined by the amount of current flowing through the resistor on its way to the capacitor. At time t_0 the current flowing to the capacitor is maximum. Thus, the voltage drop across the resistor is maximum (E = IR). As time progresses toward time t_1, the current flowing to the capacitor steadily decreases and causes the voltage developed across the resistor (R) to steadily decrease. When time t_1 is reached, current flowing to the capacitor is stopped and the voltage developed across the resistor has decreased to zero.

You should remember that capacitance opposes a change in voltage. This is shown by comparing graph (A) to graph (D). In graph (A) the voltage changed instantly from 0 volts to 6 volts across the circuit, while the voltage developed across the capacitor in graph (D) took the entire time interval from time t_0 to time t_1 to reach 6 volts. The reason for this is that in the first instant at time t_0, maximum current flows through R and the entire circuit voltage is dropped across the resistor. The voltage impressed across the capacitor at t_0 is zero volts. As time progresses toward t_1, the decreasing current causes progressively less voltage to be dropped across the resistor (R), and more voltage builds up across the capacitor (C). At time t_1, the voltage felt across the capacitor is equal to the source voltage (6 volts), and the voltage dropped across the resistor (R) is equal to zero. This is the complete charge cycle of the capacitor.

As you may have noticed, the processes which take place in the time interval t_0 to t_1 in a series RC circuit are exactly opposite to those in a series LR circuit.

For your comparison, the important points of the charge cycle of RC and LR circuits are summarized in table 3-1.

Table 3-1.—Summary of Capacitive and Inductive Characteristics

		TIME ZERO (t_0)	TIME BETWEEN t_0 AND t_1	TIME ONE (t_1)
CIRCUIT CURRENT	(R-C)	MAXIMUM	DECREASING	ZERO
	(R-L)	ZERO	INCREASING	MAXIMUM
VOLTAGE DEVELOPED ACROSS THE RESISTOR	(R-C)	MAXIMUM	DECREASING	ZERO
	(R-L)	ZERO	INCREASING	MAXIMUM
VOLTAGE DEVELOPED ACROSS CAPACITOR/ INDUCTOR	(R-C)	ZERO	INCREASING	MAXIMUM
	(R-L)	MAXIMUM	DECREASING	ZERO

DISCHARGE CYCLE

In figure 3-10 at time t_0, the capacitor is fully charged. When S1 is open and S2 closes, the capacitor discharge cycle starts. At the first instant, circuit voltage attempts to go from source potential (6 volts) to zero volts, as shown in graph (A). Remember, though, the capacitor during the charge cycle has stored energy in an electrostatic field.

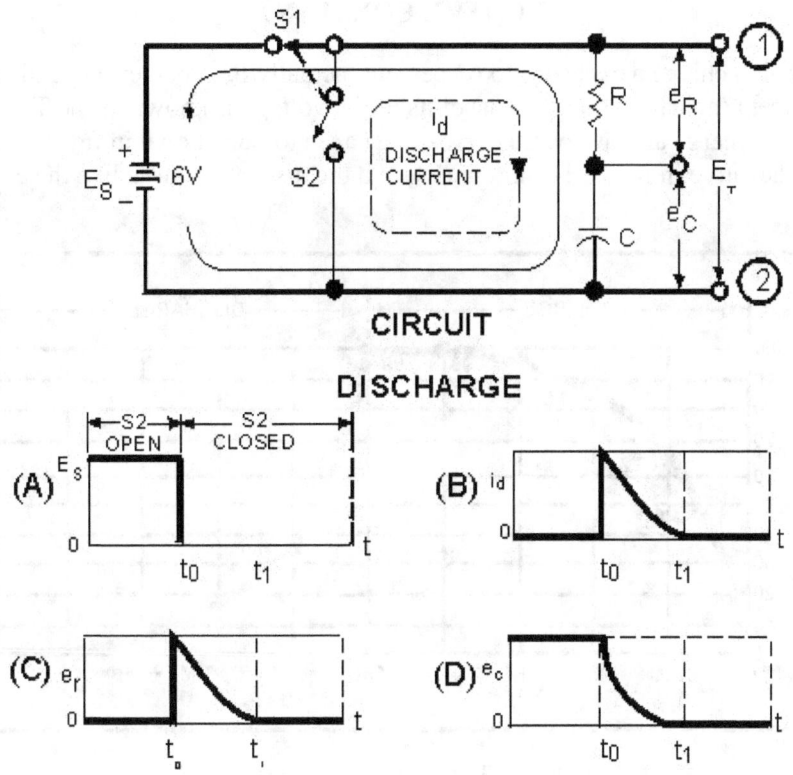

Figure 3-10.—Discharge of an RC Series circuit.

Because S2 is closed at the same time S1 is open, the stored energy of the capacitor now has a path for current to flow. At t_0, discharge current (i_d) from the bottom plate of the capacitor through the resistor (R) to the top plate of the capacitor (C) is maximum. As time progresses toward t_1, the discharge current steadily decreases until at time t_1 it reaches zero, as shown in graph (B).

The discharge causes a corresponding voltage drop across the resistor as shown in graph (C). At time t_0, the current through the resistor is maximum and the voltage drop (e_r) across the resistor is maximum. As the current through the resistor decreases, the voltage drop across the resistor decreases until at t_1 it has reached a value of zero. Graph (D) shows the voltage across the capacitor (e_c) during the discharge cycle. At time t_0 the voltage is maximum and as time progresses toward time t_1, the energy stored in the capacitor is depleted. At the same time the voltage across the resistor is decreasing, the voltage (e) across the capacitor is decreasing until at time t_1 the voltage (e_c) reaches zero.

By comparing graph (A) with graph (D) of figure 3-10, you can see the effect that capacitance has on a change in voltage. If the circuit had not contained a capacitor, the voltage would have ceased at the instant S1 was opened at time t_0. Because the capacitor is in the circuit, voltage is applied to the circuit until the capacitor has discharged completely at t_1. The effect of capacitance has been to oppose this change in voltage.

Q12. At what instant does the greatest voltage appear across the resistor in a series RC circuit when the capacitor is charging?

Q13. What is the voltage drop across the resistor in an RC charging circuit when the charge on the capacitor is equal to the battery voltage?

RC TIME CONSTANT

The time required to charge a capacitor to 63 percent (actually 63.2 percent) of full charge or to discharge it to 37 percent (actually 36.8 percent) of its initial voltage is known as the TIME CONSTANT (TC) of the circuit. The charge and discharge curves of a capacitor are shown in figure 3-11. Note that the charge curve is like the curve in figure 3-9, graph (D), and the discharge curve like the curve in figure 3-9, graph (B).

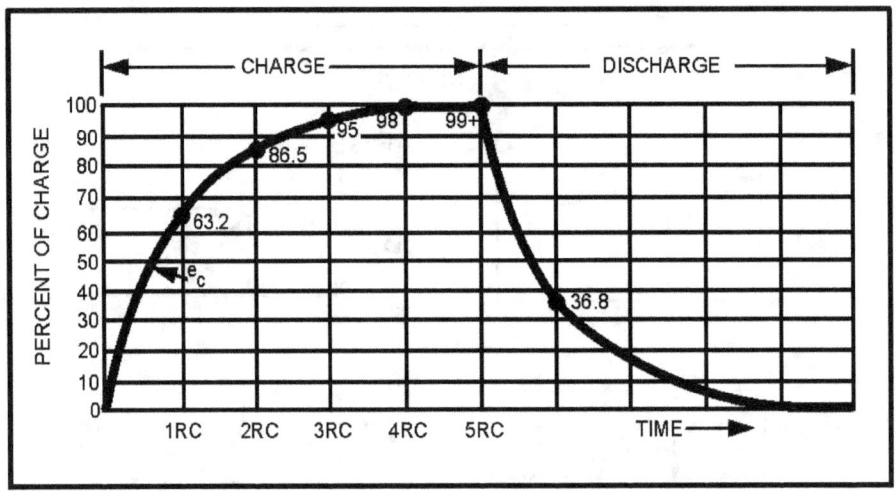

Figure 3-11.—RC time constant.

The value of the time constant in seconds is equal to the product of the circuit resistance in ohms and the circuit capacitance in farads. The value of one time constant is expressed mathematically as t = RC. Some forms of this formula used in calculating RC time constants are:

$$t \text{ (in seconds)} = R \text{ (in ohms)} \times C \text{ (in farads)}$$

$$t \text{ (in seconds)} = R \text{ (in megohms)} \times C \text{ (in microfarads)}$$

$$t \text{ (in microseconds)} = R \text{ (in ohms)} \times C \text{ (in microfarads)}$$

$$t \text{ (in microseconds)} = R \text{ (in megohms)} \times C \text{ (in picofarads)}$$

Q14. *What is the RC time constant of a series RC circuit that contains a 12-megohm resistor and a 12-microfarad capacitor?*

UNIVERSAL TIME CONSTANT CHART

Because the impressed voltage and the values of R and C or R and L in a circuit are usually known, a UNIVERSAL TIME CONSTANT CHART (fig. 3-12) can be used to find the time constant of the circuit. Curve A is a plot of both capacitor voltage during charge and inductor current during growth. Curve B is a plot of both capacitor voltage during discharge and inductor current during decay.

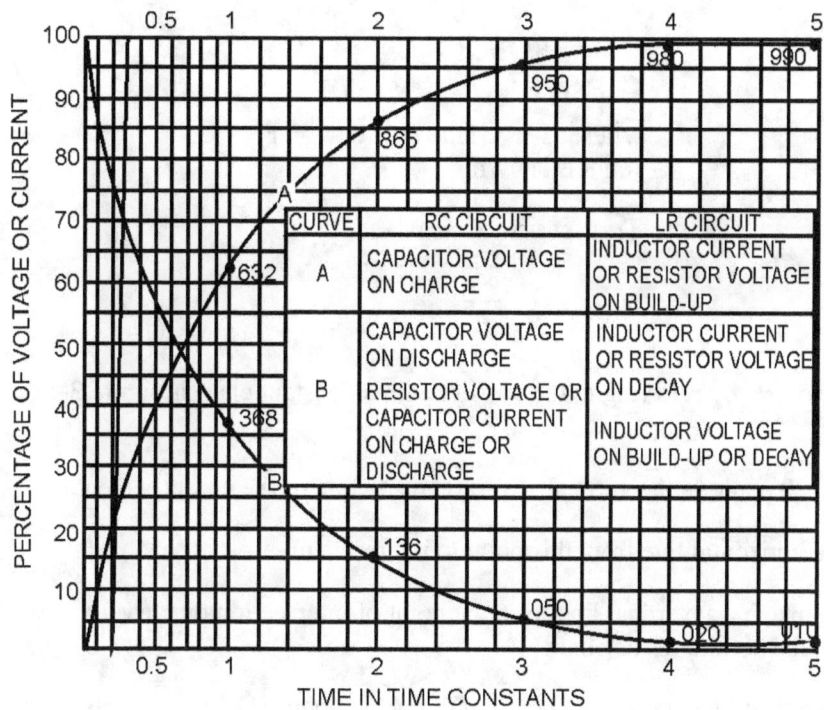

Figure 3-12.—Universal time constant chart for RC and RL circuit.

The time scale (horizontal scale) is graduated in terms of the RC or L/R time constants so that the curves may be used for any value of R and C or L and R. The voltage and current scales (vertical scales) are graduated in terms of percentage of the maximum voltage or current so that the curves may be used for any value of voltage or current. If the time constant and the initial or final voltage for the circuit in question are known, the voltages across the various parts of the circuit can be obtained from the curves for any time after the switch is closed, either on charge or discharge. The same reasoning is true of the current in the circuit.

The following problem illustrates how the universal time constant chart may be used.

An RC circuit is to be designed in which a capacitor (C) must charge to 20 percent (0.20) of the maximum charging voltage in 100 microseconds (0.0001 second). Because of other considerations, the resistor (R) must have a value of 20,000 ohms. What value of capacitance is needed?

$$\text{Given:} \quad \text{Percent of charge} = 20\% \,(.20)$$
$$t = 100 \,\mu s$$
$$R = 20{,}000 \,\Omega$$

Find: The capacitance of capacitor C.

Solution: Because the only values given are in units of time and resistance, a variation of the formula to find RC time is used:

3-17

$$RC = R \times C$$

where: 1 RC time constant = R × C
and R is known.

Transpose the formula to:

$$C = \frac{RC}{R}$$

Find the value of RC by referring to the universal time constant chart in figure 3-12 and proceed as follows:

- Locate the 20 point on the vertical scale at the left side of the chart (percentage).

- Follow the horizontal line from this point to intersect curve A.

- Follow an imaginary vertical line from the point of intersection on curve A downward to cross the RC scale at the bottom of the chart.

Note that the vertical line crosses the horizontal scale at about .22 RC as illustrated below:

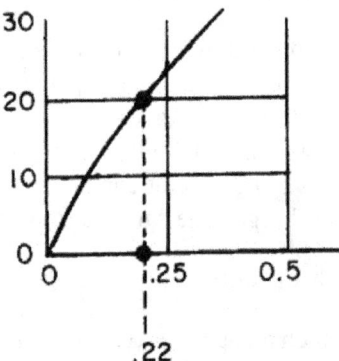

The value selected from the graph means that a capacitor (including the one you are solving for) will reach twenty percent of full charge in twenty-two one hundredths (.22) of one RC time constant. Remember that it takes 100 µs for the capacitor to reach 20% of full charge. Since 100 µs is equal to .22 RC (twenty-two one-hundredths), then the time required to reach one RC time constant must be equal to:

$$.22\,RC = 100\,\mu s$$

$$RC = \frac{1}{.22} \times 100\,\mu s$$

$$RC = \frac{100\,\mu s}{.22}$$

$$RC = 454.54\,\mu s \text{ (rounded off to } 455\,\mu s\text{)}$$

$$RC = 455\,\mu s$$

Now use the following formula to find C:

$$C = \frac{RC}{R}$$

$$C = \frac{455 \, \mu s}{20{,}000 \text{ ohms}}$$

$$C = 0.0227 \, \mu F$$

$$C = .023 \, \mu F$$

To summarize the above procedures, the problem and solution are shown below without the step by step explanation.

Given: Percent of charge = 20% (.20)
t = 100 μs
R = 20,000 ohms

Transpose the RC time constant formula as follows:

$$R \times C = RC$$

$$C = \frac{RC}{R}$$

Find: RC

$$.22 \, RC = 100 \, \mu s$$

$$RC = \frac{100 \, \mu s}{.22}$$

$$RC = 455 \, \mu s$$

Substitute the R and RC values into the formula:

$$C = \frac{RC}{R}$$

$$C = \frac{455 \, \mu s}{20{,}000} \text{ ohms}$$

$$C = .023 \, \mu s$$

The graphs shown in figure 3-11 and 3-12 are not entirely complete. That is, the charge or discharge (or the growth or decay) is not quite complete in 5 RC or 5 L/R time constants. However, when the values reach 0.99 of the maximum (corresponding to 5 RC or 5 L/R), the graphs may be considered accurate enough for all practical purposes.

Q15. A circuit is to be designed in which a capacitor must charge to 40 percent of the maximum charging voltage in 200 microseconds. The resistor to be used has a resistance of 40,000 ohms. What size capacitor must be used? (Use the universal time constant chart in figure 3-12.)

CAPACITORS IN SERIES AND PARALLEL

Capacitors may be connected in series or in parallel to obtain a resultant value which may be either the sum of the individual values (in parallel) or a value less than that of the smallest capacitance (in series).

CAPACITORS IN SERIES

The overall effect of connecting capacitors in series is to move the plates of the capacitors further apart. This is shown in figure 3-13. Notice that the junction between C1 and C2 has both a negative and a positive charge. This causes the junction to be essentially neutral. The total capacitance of the circuit is developed between the left plate of C1 and the right plate of C2. Because these plates are farther apart, the total value of the capacitance in the circuit is decreased. Solving for the total capacitance (C_T) of capacitors connected in series is similar to solving for the total resistance (R_T) of resistors connected in parallel.

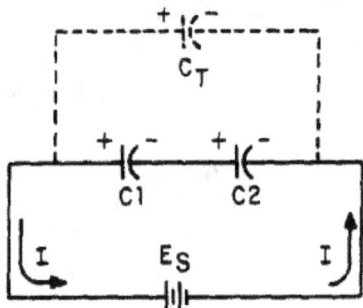

Figure 3-13.—Capacitors in series.

Note the similarity between the formulas for R_T and C_T:

$$R_T = \frac{1}{\frac{1}{R1} + \frac{1}{R2} + \ldots \frac{1}{R_n}}$$

$$C_T = \frac{1}{\frac{1}{C1} + \frac{1}{C2} + \ldots \frac{1}{C_n}}$$

If the circuit contains more than two capacitors, use the above formula. If the circuit contains only two capacitors, use the below formula:

$$C_T = \frac{C1 \times C2}{C1 + C2}$$

Note: All values for C_T, C1, C2, C3,... C_n should be in farads. It should be evident from the above formulas that the total capacitance of capacitors in series is less than the capacitance of any of the individual capacitors.

Example: Determine the total capacitance of a series circuit containing three capacitors whose values are 0.01 µF, 0.25 µF, and 50,000 pF, respectively.

Given:
$C_1 = 0.01 \mu F$
$C_2 = 0.25 \mu F$
$C_3 = 50,000 pF$

Solution:

$$C_T = \cfrac{1}{\cfrac{1}{C_1} + \cfrac{1}{C_2} + \cfrac{1}{C_3}}$$

$$C_T = \cfrac{1}{\cfrac{1}{.01 \mu F} + \cfrac{1}{.25 \mu F} + \cfrac{1}{50,000 pF}}$$

$$C_T = \cfrac{1}{\cfrac{1}{1 \times 10^{-8}} + \cfrac{1}{25 \times 10^{-8}} + \cfrac{1}{5 \times 10^{-8}}} F$$

$$C_T = \cfrac{1}{100 \times 10^6 + 4 \times 10^6 + 20 \times 10^6} F$$

$$C_T = \cfrac{1}{124 \times 10^6} F$$

$$C_T = 0.008 \mu F$$

The total capacitance of 0.008µF is slightly smaller than the smallest capacitor (0.01µF).

CAPACITORS IN PARALLEL

When capacitors are connected in parallel, one plate of each capacitor is connected directly to one terminal of the source, while the other plate of each capacitor is connected to the other terminal of the source. Figure 3-14 shows all the negative plates of the capacitors connected together, and all the positive plates connected together. C_T, therefore, appears as a capacitor with a plate area equal to the sum of all the individual plate areas. As previously mentioned, capacitance is a direct function of plate area. Connecting capacitors in parallel effectively increases plate area and thereby increases total capacitance.

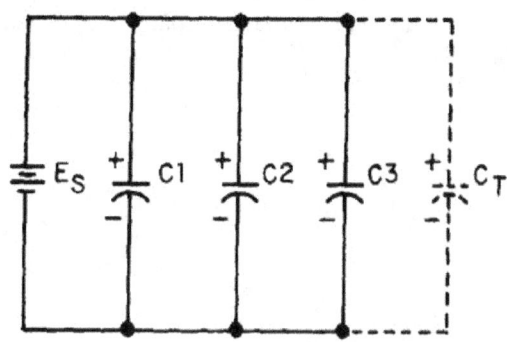

Figure 3-14.—Parallel capacitive circuit.

For capacitors connected in parallel the total capacitance is the sum of all the individual capacitances. The total capacitance of the circuit may by calculated using the formula:

$$C_T = C1 + C2 + C3 + \ldots\ldots C_n$$

where all capacitances are in the same units.

Example: Determine the total capacitance in a parallel capacitive circuit containing three capacitors whose values are 0.03 µF, 2.0 µF, and 0.25 µF, respectively.

$$\text{Given:} \quad C1 = 0.03\,\mu F$$
$$C2 = 2\,\mu F$$
$$C3 = 0.25\,\mu F$$
$$\text{Solution:} \quad C_T = C1 + C2 + C3$$
$$C_T = 0.03\,\mu F + 2.0\,\mu F + 0.25\,\mu F$$
$$C_T = 2.28\,\mu F$$

Q16. What is the total capacitance of a circuit that contains two capacitors (10 µF and 0.1 µF) wired together in series?

Q17. What is the total capacitance of a circuit in which four capacitors (10 µF, 21 µF, 0.1 µF and 2 µF) are connected in parallel?

FIXED CAPACITOR

A fixed capacitor is constructed in such manner that it possesses a fixed value of capacitance which cannot be adjusted. A fixed capacitor is classified according to the type of material used as its dielectric, such as paper, oil, mica, or electrolyte.

A PAPER CAPACITOR is made of flat thin strips of metal foil conductors that are separated by waxed paper (the dielectric material). Paper capacitors usually range in value from about 300 picofarads to about 4 microfarads. The working voltage of a paper capacitor rarely exceeds 600 volts. Paper capacitors are sealed with wax to prevent the harmful effects of moisture and to prevent corrosion and leakage.

Many different kinds of outer covering are used on paper capacitors, the simplest being a tubular cardboard covering. Some types of paper capacitors are encased in very hard plastic. These types are very rugged and can be used over a much wider temperature range than can the tubular cardboard type. Figure 3-15(A) shows the construction of a tubular paper capacitor; part 3-15(B) shows a completed cardboard-encased capacitor.

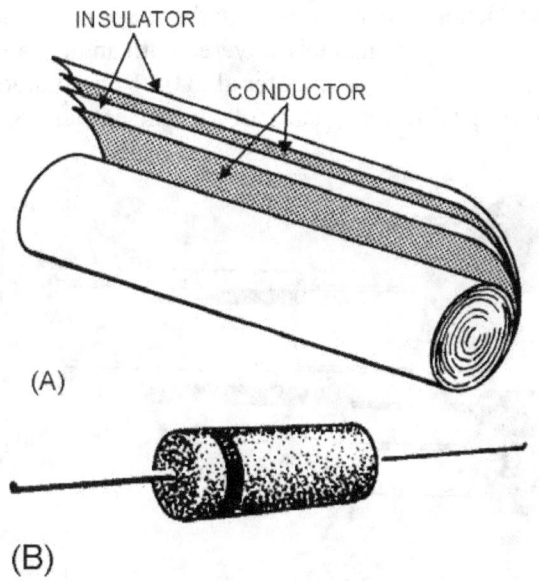

Figure 3-15.—Paper capacitor.

A MICA CAPACITOR is made of metal foil plates that are separated by sheets of mica (the dielectric). The whole assembly is encased in molded plastic. Figure 3-16(A) shows a cut-away view of a mica capacitor. Because the capacitor parts are molded into a plastic case, corrosion and damage to the plates and dielectric are prevented. In addition, the molded plastic case makes the capacitor mechanically stronger. Various types of terminals are used on mica capacitors to connect them into circuits. These terminals are also molded into the plastic case.

Mica is an excellent dielectric and can withstand a higher voltage than can a paper dielectric of the same thickness. Common values of mica capacitors range from approximately 50 picofarads to 0.02 microfarad. Some different shapes of mica capacitors are shown in figure 3-16(B).

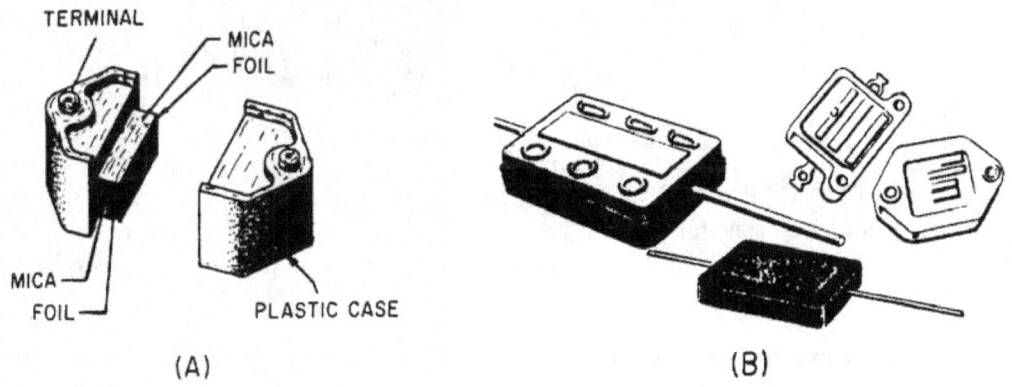

Figure 3-16.—Typical mica capacitors.

A CERAMIC CAPACITOR is so named because it contains a ceramic dielectric. One type of ceramic capacitor uses a hollow ceramic cylinder as both the form on which to construct the capacitor and as the dielectric material. The plates consist of thin films of metal deposited on the ceramic cylinder.

A second type of ceramic capacitor is manufactured in the shape of a disk. After leads are attached to each side of the capacitor, the capacitor is completely covered with an insulating moisture-proof coating. Ceramic capacitors usually range in value from 1 picofarad to 0.01 microfarad and may be used with voltages as high as 30,000 volts. Some different shapes of ceramic capacitors are shown in figure 3-17.

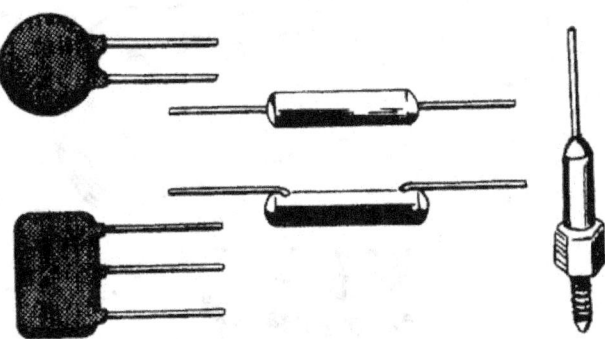

Figure 3-17.—Ceramic capacitors.

Examples of ceramic capacitors.

An ELECTROLYTIC CAPACITOR is used where a large amount of capacitance is required. As the name implies, an electrolytic capacitor contains an electrolyte. This electrolyte can be in the form of a liquid (wet electrolytic capacitor). The wet electrolytic capacitor is no longer in popular use due to the care needed to prevent spilling of the electrolyte.

A dry electrolytic capacitor consists essentially of two metal plates separated by the electrolyte. In most cases the capacitor is housed in a cylindrical aluminum container which acts as the negative terminal of the capacitor (see fig. 3-18). The positive terminal (or terminals if the capacitor is of the multisection type) is a lug (or lugs) on the bottom end of the container. The capacitance value(s) and the voltage rating of the capacitor are generally printed on the side of the aluminum case.

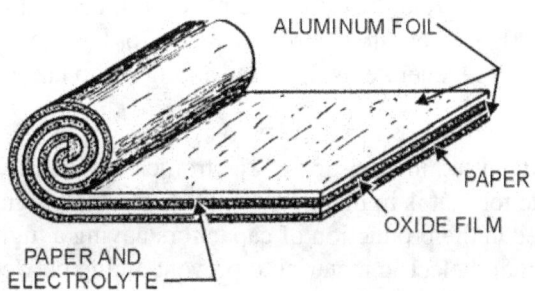

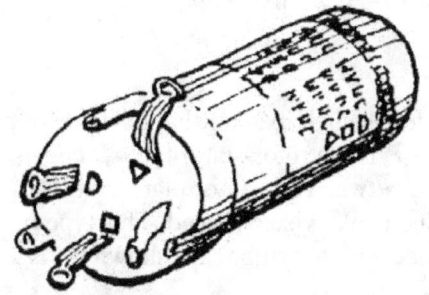

Figure 3-18.—Construction of an electrolytic capacitor.

An example of a multisection electrolytic capacitor is illustrated in figure 3-18(B). The four lugs at the end of the cylindrical aluminum container indicates that four electrolytic capacitors are enclosed in the can. Each section of the capacitor is electrically independent of the other sections. It is possible for one section to be defective while the other sections are still good. The can is the common negative connection to the four capacitors. Separate terminals are provided for the positive plates of the capacitors. Each capacitor is identified by an embossed mark adjacent to the lugs, as shown in figure 3-18(B). Note the identifying marks used on the electrolytic capacitor are the half moon, the triangle, the square, and no embossed mark. By looking at the bottom of the container and the identifying sheet pasted to the side of the container, you can easily identify the value of each section.

Internally, the electrolytic capacitor is constructed similarly to the paper capacitor. The positive plate consists of aluminum foil covered with an extremely thin film of oxide. This thin oxide film (which is formed by an electrochemical process) acts as the dielectric of the capacitor. Next to and in contact with the oxide is a strip of paper or gauze which has been impregnated with a paste-like electrolyte. The electrolyte acts as the negative plate of the capacitor. A second strip of aluminum foil is then placed against the electrolyte to provide electrical contact to the negative electrode (the electrolyte). When the three layers are in place they are rolled up into a cylinder as shown in figure 3-18(A).

An electrolytic capacitor has two primary disadvantages compared to a paper capacitor in that the electrolytic type is POLARIZED and has a LOW LEAKAGE RESISTANCE. This means that should the positive plate be accidentally connected to the negative terminal of the source, the thin oxide film dielectric will dissolve and the capacitor will become a conductor (i.e., it will short). The polarity of the terminals is normally marked on the case of the capacitor. Since an electrolytic capacitor is polarity sensitive, its use is ordinarily restricted to a dc circuit or to a circuit where a small ac voltage is superimposed on a dc voltage. Special electrolytic capacitors are available for certain ac applications, such as a motor starting capacitor. Dry electrolytic capacitors vary in size from about 4 microfarads to several thousand microfarads and have a working voltage of approximately 500 volts.

The type of dielectric used and its thickness govern the amount of voltage that can safely be applied to the electrolytic capacitor. If the voltage applied to the capacitor is high enough to cause the atoms of the dielectric material to become ionized, arcing between the plates will occur. In most other types of capacitors, arcing will destroy the capacitor. However, an electrolytic capacitor has the ability to be self-healing. If the arcing is small, the electrolytic will regenerate itself. If the arcing is too large, the capacitor will not self-heal and will become defective.

OIL CAPACITORS are often used in high-power electronic equipment. An oil-filled capacitor is nothing more than a paper capacitor that is immersed in oil. Since oil impregnated paper has a high dielectric constant, it can be used in the production of capacitors having a high capacitance value. Many capacitors will use oil with another dielectric material to prevent arcing between the plates. If arcing should occur between the plates of an oil-filled capacitor, the oil will tend to reseal the hole caused by the arcing. Such a capacitor is referred to as a SELF-HEALING capacitor.

VARIABLE CAPACITOR

A variable capacitor is constructed in such manner that its value of capacitance can be varied. A typical variable capacitor (adjustable capacitor) is the rotor-stator type. It consists of two sets of metal plates arranged so that the rotor plates move between the stator plates. Air is the dielectric. As the position of the rotor is changed, the capacitance value is likewise changed. This type of capacitor is used for tuning most radio receivers. Its physical appearance and its symbol are shown in figure 3-19.

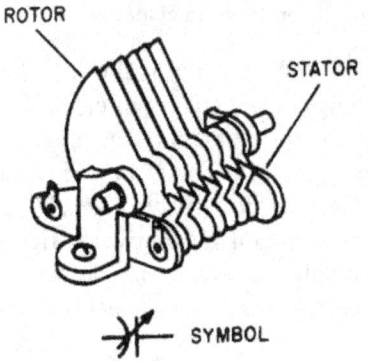

Figure 3-19.—Rotor-stator type variable capacitor.

Another type of variable capacitor (trimmer capacitor) and its symbol are shown in figure 3-20. This capacitor consists of two plates separated by a sheet of mica. A screw adjustment is used to vary the distance between the plates, thereby changing the capacitance.

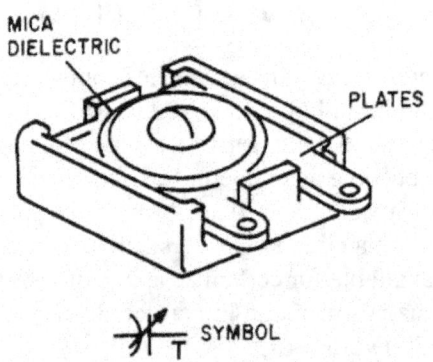

Figure 3-20.—Trimmer capacitor.

Q18.

    a. An oxide-film dielectric is used in what type of capacitor?

    b. A screw adjustment is used to vary the distance between the plates of what type of capacitor?

COLOR CODES FOR CAPACITORS

Although the capacitance value may be printed on the body of a capacitor, it may also be indicated by a color code. The color code used to represent capacitance values is similar to that used to represent resistance values. The color codes currently in use are the Joint Army-Navy (JAN) code and the Radio Manufacturers' Association (RMA) code.

For each of these codes, colored dots or bands are used to indicate the value of the capacitor. A mica capacitor, it should be noted, may be marked with either three dots or six dots. Both the three- and the six-dot codes are similar, but the six-dot code contains more information about electrical ratings of the capacitor, such as working voltage and temperature coefficient.

The capacitor shown in figure 3-21 represents either a mica capacitor or a molded paper capacitor. To determine the type and value of the capacitor, hold the capacitor so that the three arrows point left to right (>). The first dot at the base of the arrow sequence (the left-most dot) represents the capacitor TYPE. This dot is either black, white, silver, or the same color as the capacitor body. Mica is represented by a black or white dot and paper by a silver dot or dot having the same color as the body of the capacitor. The two dots to the immediate right of the type dot indicate the first and second digits of the capacitance value. The dot at the bottom right represents the multiplier to be used. The multiplier represents picofarads. The dot in the bottom center indicates the tolerance value of the capacitor.

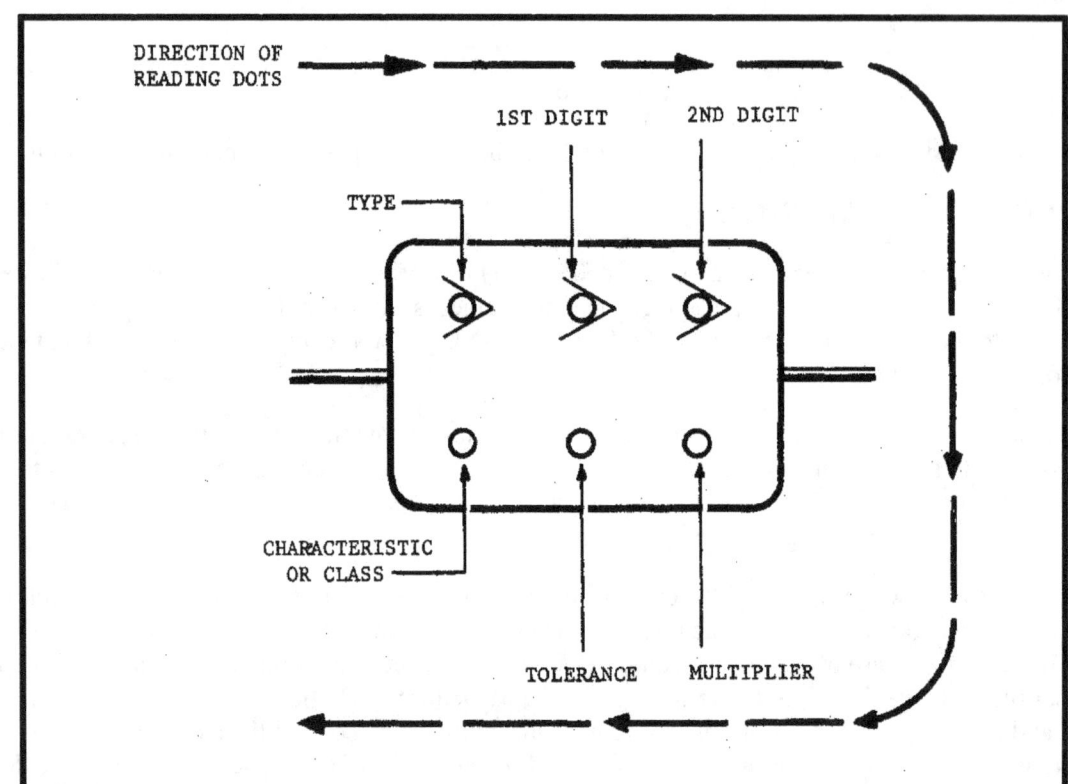

TYPE	COLOR	1ST DIGIT	2ND DIGIT	MULTIPLIER	TOLERANCE (PERCENT)	CHARACTERISTIC OR CLASS
JAN, MICA	BLACK	0	0	1.0	±20	APPLIES TO
	BROWN	1	1	10		TEMPERATURE
	RED	2	2	100	± 2	COEFFICIENT
	ORANGE	3	3	1,000	± 3	OR METHODS
	YELLOW	4	4	10,000	± 4	OF TESTING
	GREEN	5	5	100,000	± 5	
	BLUE	6	6	1,000,000	± 6	
	VIOLET	7	7	10,000,000	± 7	
	GRAY	8	8	100,000,000	± 8	
EIA, MICA	WHITE	9	9	1,000,000,000	+ 9	
	GOLD			.1		
MOLDED PAPER	SILVER			.01	±10	
	BODY				±20	

Figure 3-21.—6-dot color code for mica and molded paper capacitors.

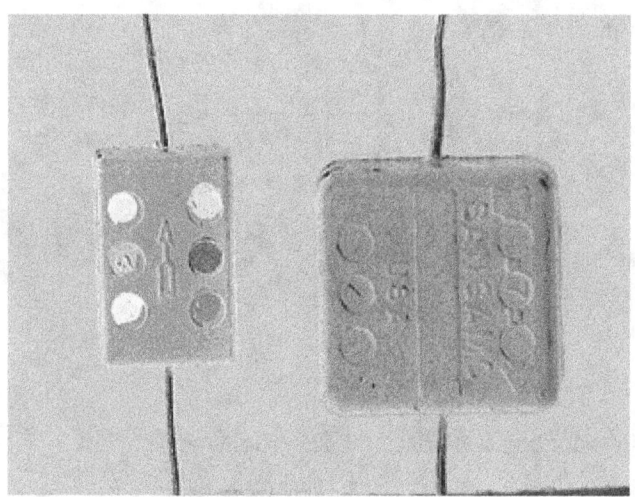

Example of mica capacitors.

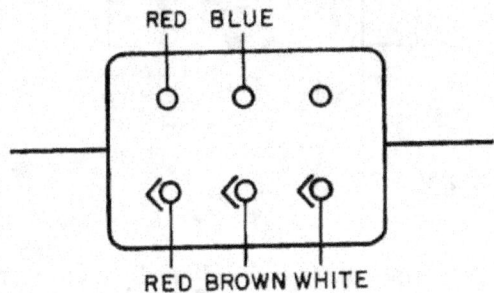

Example of mica capacitors.

To read the capacitor color code on the above capacitor:

1. Hold the capacitor so the arrows point left to right.

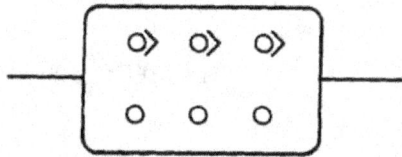

2. Read the first dot.

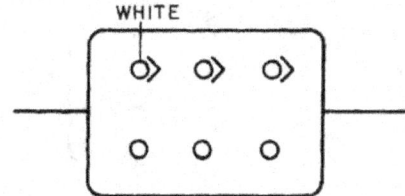

White = mica

3-29

3. Read the first digit dot.

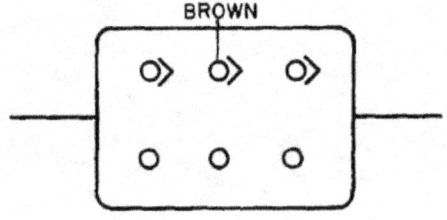

Brown = 1

4. Read the second digit dot and apply it to the first digit.

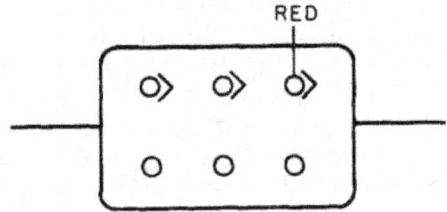

Red = 2 → 12

5. Read the multiplier dot and multiply the first two digits by multiplier. (Remember that the multiplier is in picofarads).

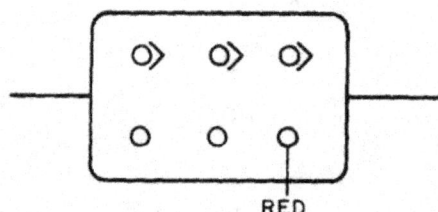

Red = 100 → 12 x 100 = 1200 pF

6. Lastly, read the tolerance dot.

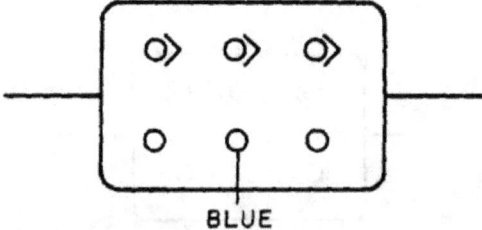

Blue = ±6%

3-30

According to the above coding, the capacitor is a mica capacitor whose capacitance is 1200 pF with a tolerance of ±6%.

The capacitor shown in figure 3-22 is a tubular capacitor. Because this type of capacitor always has a paper dielectric, the type code is omitted. To read the code, hold the capacitor so the band closest to the end is on the left side; then read left to right. The last two bands (the fifth and sixth bands from the left) represent the voltage rating of the capacitor. This means that if a capacitor is coded red, red, red, yellow, yellow, yellow, it has the following digit values:

red	=	2
red	=	2
red	=	× 100 pF
yellow	=	±40%
yellow	=	4
yellow	=	4

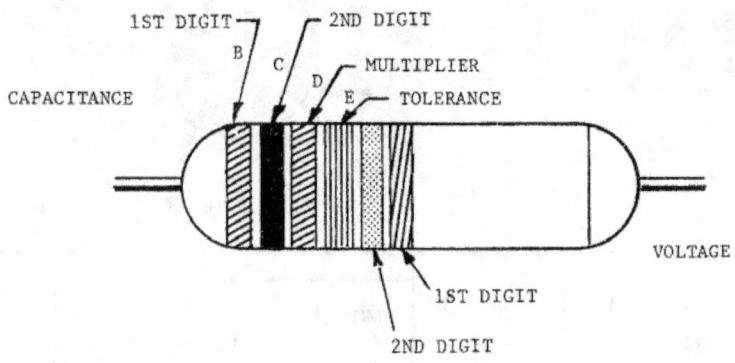

COLOR	CAPACITANCE			TOLERANCE (PERCENT)	VOLTAGE RATING	
	1ST DIGIT	2ND DIGIT	MULTIPLIER		1ST DIGIT	2ND DIGIT
BLACK	0	0	1	±20	0	0
BROWN	1	1	10		1	1
RED	2	2	100		2	2
ORANGE	3	3	1,000	±30	3	3
YELLOW	4	4	10,000	±40	4	4
GREEN	5	5	100,000	± 5	5	5
BLUE	6	6	1,000,000		6	6
VIOLET	7	7			7	7
GRAY	8	8			8	8
WHITE	9	9		±10	9	9

Figure 3-22.—6-band color code for tubular paper dielectric capacitors.

The six digits indicate a capacitance of 2200 pF with a ±40 percent tolerance and a working voltage of 44 volts.

The ceramic capacitor is color coded as shown in figure 3-23 and the mica capacitor as shown in figure 3-24. Notice that this type of mica capacitor differs from the one shown in figure 3-21 in that the arrow is solid instead of broken. This type of mica capacitor is read in the same manner as the one shown in figure 3-21, with one exception: the first dot indicates the first digit. (Note: Because this type of capacitor is always mica, there is no need for a type dot.)

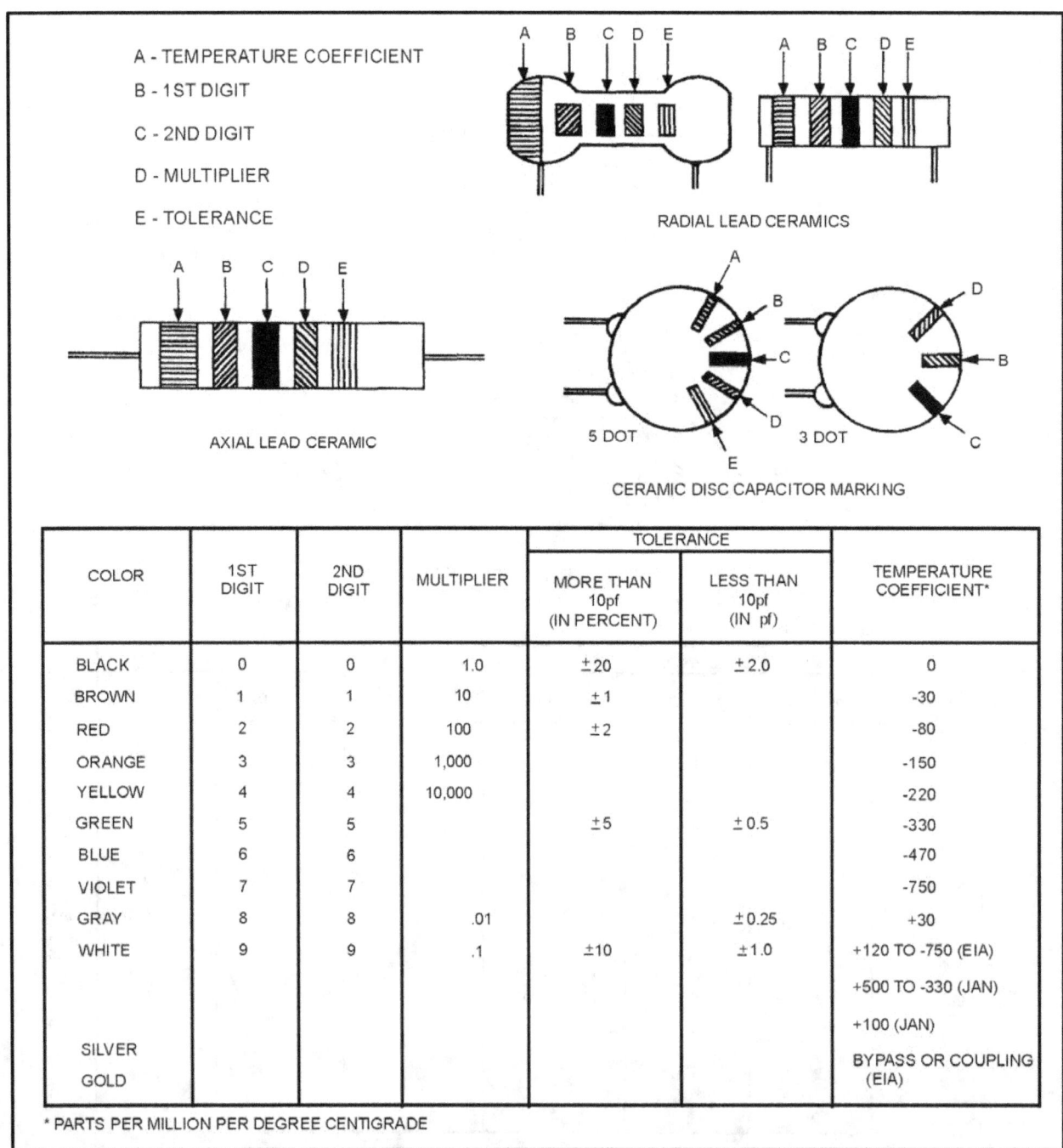

Figure 3-23.—Ceramic capacitor color code.

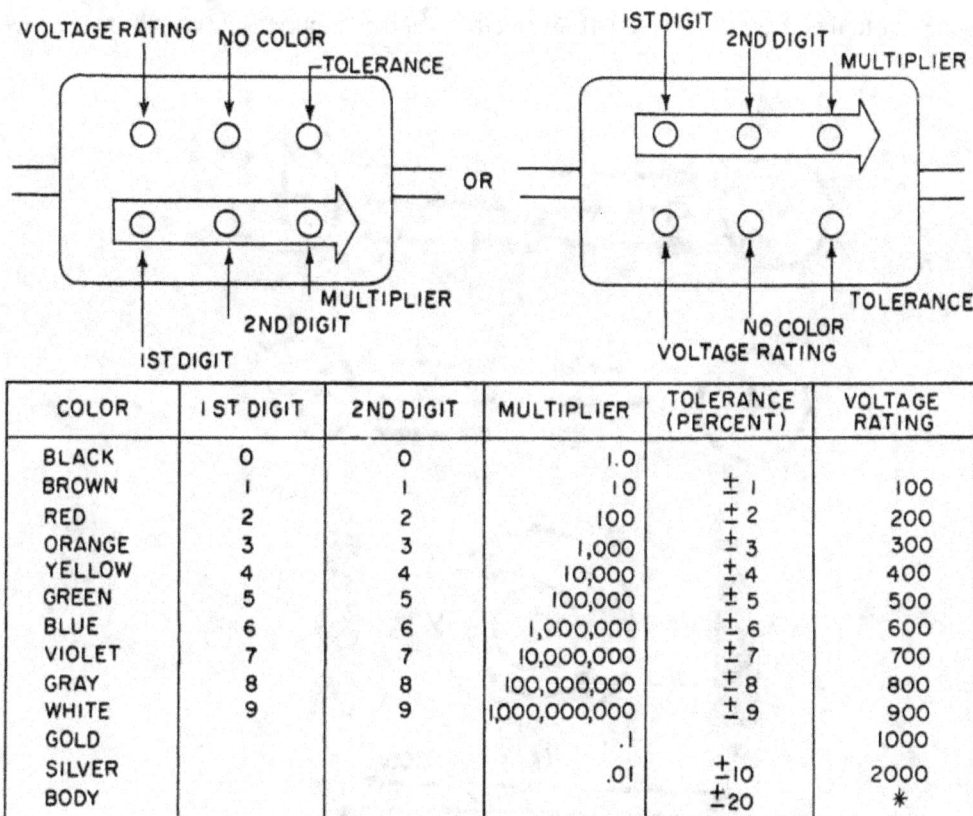

Figure 3-24.—Mica capacitor color code.

Q19. Examine the three capacitors shown below. What is the capacitance of each?

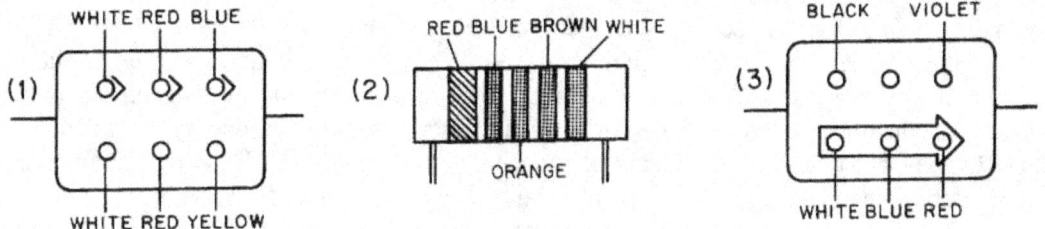

SUMMARY

Before going on to the next chapter, study the below summary to be sure that you understand the important points of this chapter.

THE ELECTROSTATIC FIELD—When a charged body is brought close to another charged body, the bodies either attract or repel one another. (If the charges are alike they repel; if the charges are opposite they attract). The field that causes this effect is called the ELECTROSTATIC FIELD. The amount by which two charges attract or repel each other depends upon the size of the charges and the distance between the charges. The electrostatic field (force between two charged bodies) may be represented by lines of force drawn perpendicular to the charged surfaces. If an electron is placed in the field, it will move toward the positive charge.

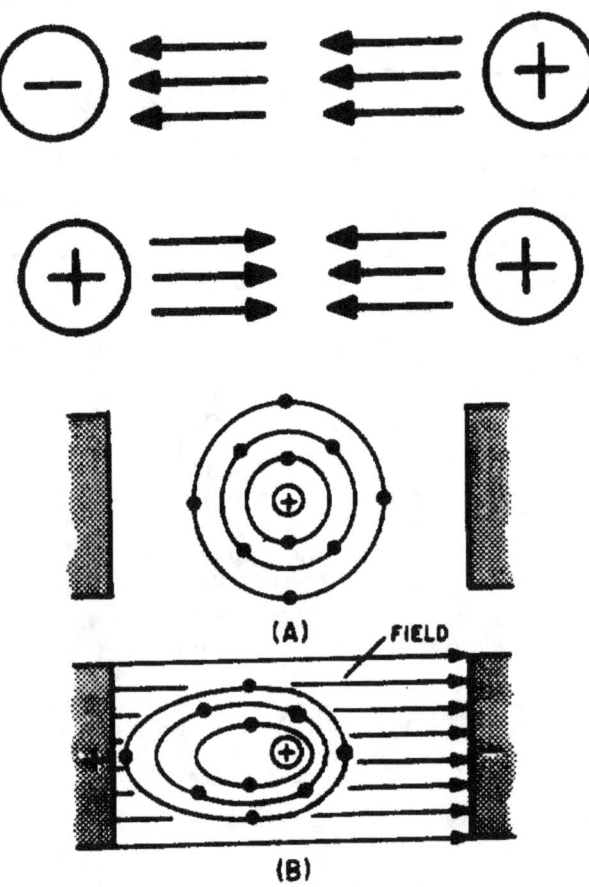

CAPACITANCE—Capacitance is the property of a circuit which OPPOSES any CHANGE in the circuit VOLTAGE. The effect of capacitance may be seen in any circuit where the voltage is changing. Capacitance is usually defined as the ability of a circuit to store electrical energy. This energy is stored in an electrostatic field. The device used in an electrical circuit to store this charge (energy) is called a CAPACITOR. The basic unit of measurement of capacitance is the FARAD (F). A one-farad capacitor will store one coulomb of charge (energy) when a potential of one volt is applied across the capacitor plates. The farad is an enormously large unit of capacitance. More practical units are the microfarad (μF) or the picofarad (pF).

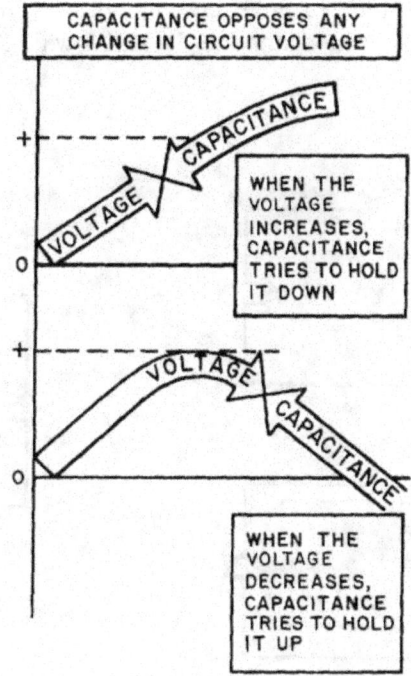

CAPACITOR—A capacitor is a physical device consisting of two pieces of conducting material separated by an insulating material. This insulating material is referred to as the DIELECTRIC. Because the dielectric is an insulator, NO current flows through the capacitor. If the dielectric breaks down and becomes a conductor, the capacitor can no longer hold a charge and is useless. The ability of a dielectric to hold a charge without breaking down is referred to as the dielectric strength. The measure of the ability of the dielectric material to store energy is called the dielectric constant. The dielectric constant is a relative value based on 1.0 for a vacuum.

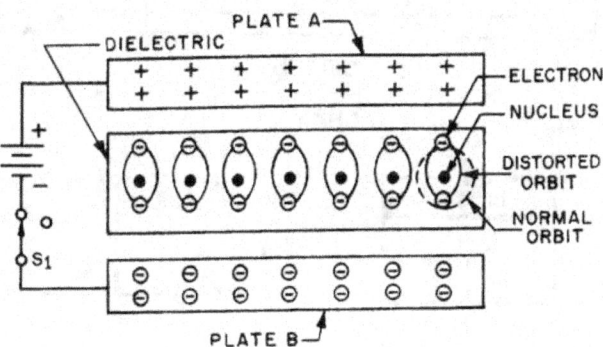

CAPACITORS IN A DC CIRCUIT—When a capacitor is connected to the terminals of a battery, each plate of the capacitor becomes charged. The plate connected to the positive terminal loses electrons. Because this plate has a lack of electrons, it assumes a positive charge. The plate connected to the negative terminal gains electrons. Because the plate has an excess of electrons, it assumes a negative charge. This process continues until the charge across the plates equals the applied voltage. At this point current ceases to flow in the circuit. As long as nothing changes in the circuit, the capacitor will hold its charge and there will be no current in any part of the circuit. If the leads of the capacitor are now shorted together, current again flows in the circuit. Current will continue to flow until the charges on the two plates become equal. At this point, current ceases to flow. With a dc voltage source, current will flow in the circuit only long enough to charge (or discharge) the capacitor. Thus, a capacitor does NOT allow dc current to flow continuously in a circuit.

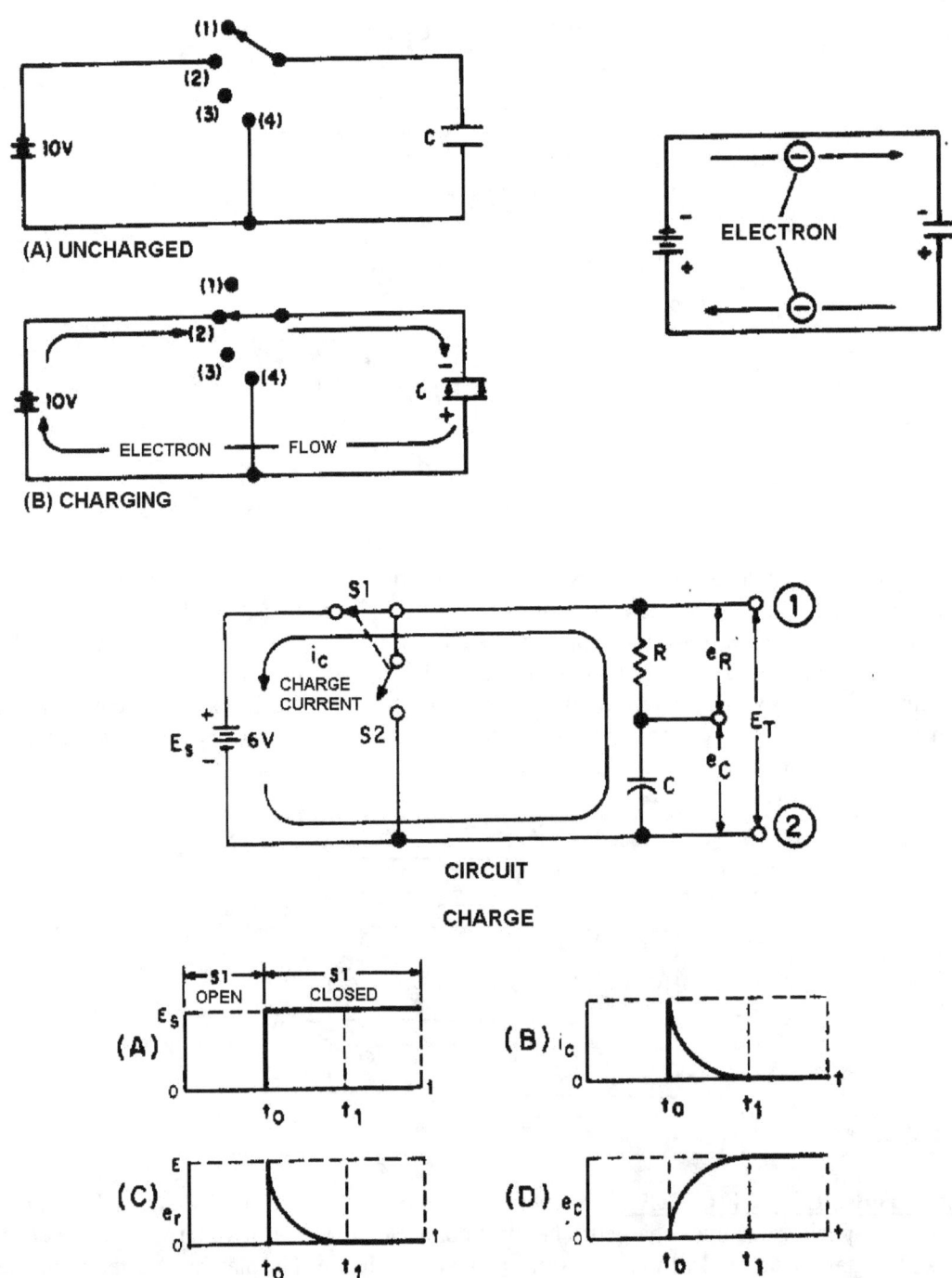

FACTORS AFFECTING CAPACITANCE—There are three factors affecting capacitance. One factor is the area of the plate surfaces. Increasing the area of the plate increases the capacitance. Another factor is the amount of space between the plates. The closer the plates, the greater will be the electrostatic field. A greater electrostatic field causes a greater capacitance. The plate spacing is determined by the thickness of the dielectric. The third factor affecting capacitance is the dielectric constant. The value of the dielectric constant is dependent upon the type of dielectric used.

WORKING VOLTAGE—The working voltage of a capacitor is the maximum voltage that can be steadily applied to the capacitor without the capacitor breaking down (shorting). The working voltage depends upon the type of material used as the dielectric (the dielectric constant) and the thickness of the dielectric.

CAPACITOR LOSSES—Power losses in a capacitor are caused by dielectric leakage and dielectric hysteresis. Dielectric leakage loss is caused by the leakage current through the resistance in the dielectric. Although this resistance is extremely high, a small amount of current does flow. Dielectric hysteresis may be defined as an effect in a dielectric material similar to the hysteresis found in a magnetic material.

RC TIME CONSTANT—The time required to charge a capacitor to 63.2 percent of the applied voltage, or to discharge the capacitor to 36.8 percent of its charge. The time constant (t) is equal to the product of the resistance and the capacitance. Expressed as a formula:

$$t = RC$$

where t is in seconds, R is in ohms, and C is in farads.

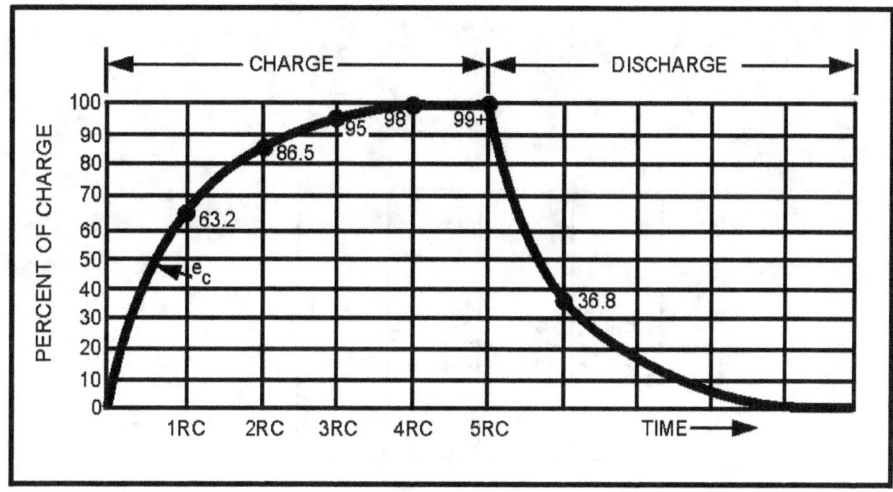

CAPACITORS IN SERIES—The effect of wiring capacitors in series is to increase the distance between plates. This reduces the total capacitance of the circuit. Total capacitance for series connected capacitors may be computed by the formula:

$$C_T = \frac{1}{\frac{1}{C1} + \frac{1}{C2} + \frac{1}{C3} + \ldots \frac{1}{C_n}}$$

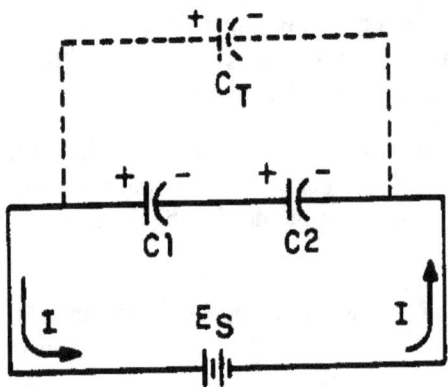

If an electrical circuit contains only two series connected capacitors, C_T may be computed using the following formula:

$$C_T = \frac{C_1 C_2}{C_1 + C_2}$$

CAPACITORS IN PARALLEL—The effect of wiring capacitors in parallel is to increase the plate area of the capacitors. Total capacitance (C_T) may be found using the formula:

$$C_T = C_1 + C_2 \ldots + C_n$$

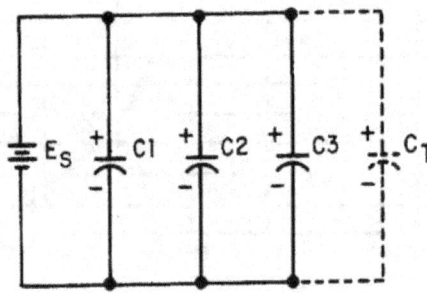

TYPES OF CAPACITORS—Capacitors are manufactured in various forms and may be divided into two main classes-fixed capacitors and variable capacitors. A fixed capacitor is constructed to have a constant or fixed value of capacitance. A variable capacitor allows the capacitance to be varied or adjusted.

ANSWERS TO QUESTIONS Q1. THROUGH Q19.

A1.
- a. A capacitor is a device that stores electrical energy in an electrostatic field.
- b. Capacitance is the property of a circuit which opposes changes in voltage.

A2.
- a. They are polarized from positive to negative.
- b. They radiate from a charged particle in straight lines and do not form closed loops.
- c. They have the ability to pass through any known material.
- d. They have the ability to distort the orbits of electrons circling the nucleus.

A3. Toward the positive charge.

A4. Two pieces of conducting material separated by an insulator.

A5. A farad is the unit of capacitance. A capacitor has a capacitance of 1 farad when a difference of 1 volt will charge it with 1 coulomb of electrons.

A6.
- a. One microfarad equals 10^{-6} farad.
- b. One picofarad equals 10^{-12} farad.

A7.
- a. The area of the plates.
- b. The distance between the plates.
- c. The dielectric constant of the material between the plates.

A8.

4372 picofarads

$$C = .2249 \left(\frac{KA}{d}\right)$$

$$C = .2249 \left(\frac{81 \times 6}{.025}\right)$$

$$C = 4372 \text{ (Rounded off)}$$

A9.
- a. Hysteresis
- b. Dielectric leakage

A10.
- a. It is the maximum voltage the capacitor can work without risk of damage.
- b. 900 volts.

A11.
- a. When the capacitor is charging, electrons accumulate on the negative plate and leave the positive plate until the charge on the capacitor is equal to the battery voltage.
- b. When the capacitor is discharging, electrons flow from the negatively charged plate to the positively charged plate until the charge on each plate is neutral.

A12. At the instant of the initiation of the action.

A13. Zero.

A14.

144 seconds
$t = R \text{ (megohms)} \times C \text{ (microfarads)}$
$t = 12 \times 12$
$t = 144 \text{ seconds}$

A15.

.01 microfarads 40% from the graph = .5
$RC = \dfrac{200}{.5}$
$RC = 400 \text{ microseconds}$
$C = \dfrac{t}{R}$
$C = \dfrac{400\,\mu s}{40{,}000\,\Omega}$
$C = .01\,\mu F = 10{,}000\,pF$

A16.

$.1\,\mu F$

$$C_T = \frac{C_1 C_2}{C_1 + C_2}$$

$$C_T = \frac{10 \times 0.1}{10 + 0.1}\,\mu F$$

$$C_T = \frac{1}{10.1}\,\mu F$$

$$C_T = .099\,\mu F \text{ or } 0.1\,\mu F$$

A17.

$33.1\,\mu F$

$$C_T = C1 + C2 + C3 + C4$$
$$C_T = 10\,\mu F + 21\,\mu F + 0.1\,\mu F + 2\,\mu F$$
$$C_T = 33.1\,\mu F$$

A18.

    a. Electrolytic capacitor

    b. Trimmer capacitor

A19.

    a. $26\,\mu F$ or $260{,}000\,pF$

    b. $630\,pF$

    c. $9600\,pF$

CHAPTER 4

INDUCTIVE AND CAPACITIVE REACTANCE

LEARNING OBJECTIVES

Upon completion of this chapter you will be able to:

1. State the effects an inductor has on a change in current and a capacitor has on a change in voltage.

2. State the phase relationships between current and voltage in an inductor and in a capacitor.

3. State the terms for the opposition an inductor and a capacitor offer to ac

4. Write the formulas for inductive and capacitive reactances.

5. State the effects of a change in frequency on X_L and X_C.

6. State the effects of a change in inductance on X_L and a change in capacitance on X_C.

7. Write the formula for determining total reactance (X); compute total reactance (X) in a series circuit; and indicate whether the total reactance is capacitive or inductive.

8. State the term given to the total opposition (Z) in an ac circuit.

9. Write the formula for impedance, and calculate the impedance in a series circuit when the values of X_C, X_L, and R are given.

10. Write the Ohm's law formulas used to determine voltage and current in an ac circuit.

11. Define true power, reactive power, and apparent power; state the unit of measurement for and the formula used to calculate each.

12. State the definition of and write the formula for power factor.

13. Given the power factor and values of X and R in an ac circuit, compute the value of reactance in the circuit, and state the type of reactance that must be connected in the circuit to correct the power factor to unity (1).

14. State the difference between calculating impedance in a series ac circuit and in a parallel ac circuit.

INDUCTIVE AND CAPACITIVE REACTANCE

You have already learned how inductance and capacitance individually behave in a direct current circuit. In this chapter you will be shown how inductance, capacitance, and resistance affect alternating current.

INDUCTANCE AND ALTERNATING CURRENT

This might be a good place to recall what you learned about phase in chapter 1. When two things are in step, going through a cycle together, falling together and rising together, they are in phase. When they are out of phase, the angle of lead or lag-the number of electrical degrees by which one of the values leads or lags the other-is a measure of the amount they are out of step. The time it takes the current in an inductor to build up to maximum and to fall to zero is important for another reason. It helps illustrate a very useful characteristic of inductive circuits-<u>the current through the inductor always lags the voltage across the inductor</u>.

A circuit having pure resistance (if such a thing were possible) would have the alternating current through it and the voltage across it rising and failing together. This is illustrated in figure 4-1(A), which shows the sine waves for current and voltage in a purely resistive circuit having an ac source. The current and voltage do not have the same amplitude, but they are in phase.

In the case of a circuit having inductance, the opposing force of the counter emf would be enough to keep the current from remaining in phase with the applied voltage. You learned that in a dc circuit containing pure inductance the current took time to rise to maximum even though the full applied voltage was immediately at maximum. Figure 4-1(B) shows the wave forms for a purely inductive ac circuit in steps of quarter-cycles.

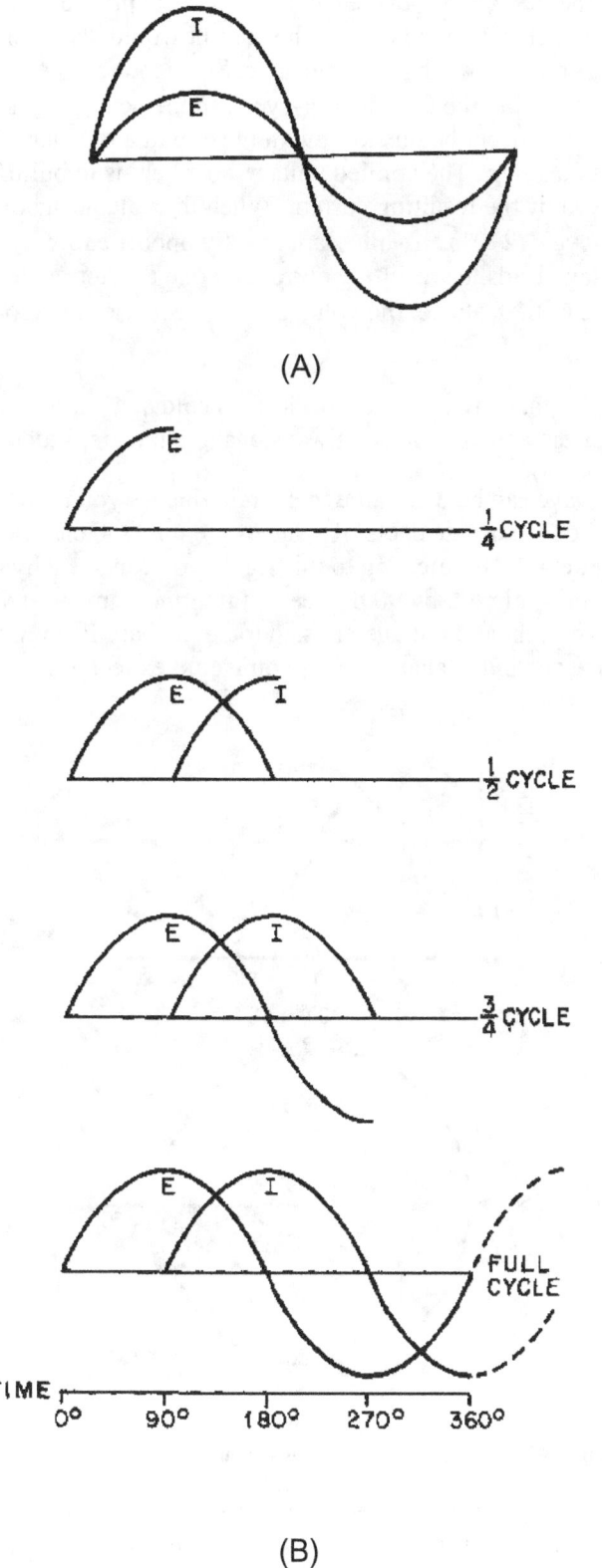

Figure 4-1.—Voltage and current waveforms in an inductive circuit.

4-3

With an ac voltage, in the first quarter-cycle (0° to 90°) the applied ac voltage is continually increasing. If there was no inductance in the circuit, the current would also increase during this first quarter-cycle. You know this circuit does have inductance. Since inductance opposes any change in current flow, no current flows during the first quarter-cycle. In the next quarter-cycle (90° to 180°) the voltage decreases back to zero; current begins to flow in the circuit and reaches a maximum value at the same instant the voltage reaches zero. The applied voltage now begins to build up to maximum in the other direction, to be followed by the resulting current. When the voltage again reaches its maximum at the end of the third quarter-cycle (270°) all values are exactly opposite to what they were during the first half-cycle. The applied voltage leads the resulting current by one quarter-cycle or 90 degrees. To complete the full 360° cycle of the voltage, the voltage again decreases to zero and the current builds to a maximum value.

You must not get the idea that any of these values stops cold at a particular instant. Until the applied voltage is removed, both current and voltage are always changing in amplitude and direction.

As you know the sine wave can be compared to a circle. Just as you mark off a circle into 360 degrees, you can mark off the time of one cycle of a sine wave into 360 electrical degrees. This relationship is shown in figure 4-2. By referring to this figure you can see why the current is said to lag the voltage, in a purely inductive circuit, by 90 degrees. Furthermore, by referring to figures 4-2 and 4-1(A) you can see why the current and voltage are said to be in phase in a purely resistive circuit. In a circuit having both resistance and inductance then, as you would expect, the current lags the voltage by an amount somewhere between 0 and 90 degrees.

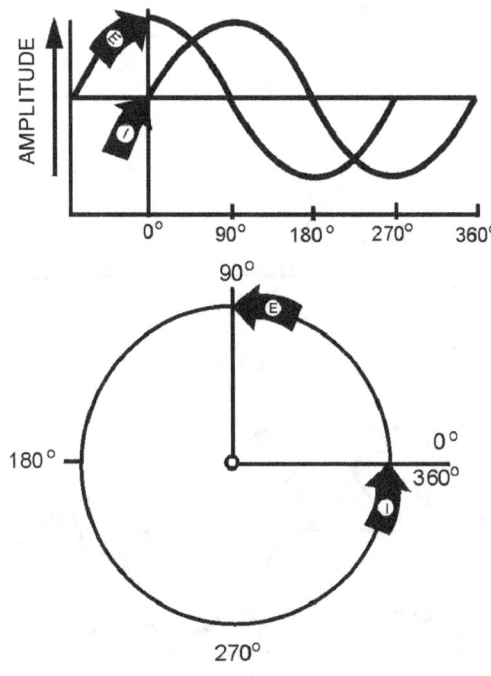

Figure 4-2.—Comparison of sine wave and circle in an inductive circuit.

A simple memory aid to help you remember the relationship of voltage and current in an inductive circuit is the word ELI. Since E is the symbol for voltage, L is the symbol for inductance, and I is the symbol for current; the word ELI demonstrates that current comes after (Lags) voltage in an inductor.

Q1. What effect does an inductor have on a change in current?

Q2. What is the phase relationship between current and voltage in an inductor?

INDUCTIVE REACTANCE

When the current flowing through an inductor continuously reverses itself, as in the case of an ac source, the inertia effect of the cemf is greater than with dc. The greater the amount of inductance (L), the greater the opposition from this inertia effect. Also, the faster the reversal of current, the greater this inertial opposition. This opposing force which an inductor presents to the FLOW of alternating current cannot be called resistance, since it is not the result of friction within a conductor. The name given to it is INDUCTIVE REACTANCE because it is the "reaction" of the inductor to the changing value of alternating current. Inductive reactance is measured in ohms and its symbol is X_L.

As you know, the induced voltage in a conductor is proportional to the rate at which magnetic lines of force cut the conductor. The greater the rate (the higher the frequency), the greater the cemf. Also, the induced voltage increases with an increase in inductance; the more ampere-turns, the greater the cemf. Reactance, then, increases with an increase of frequency and with an increase of inductance. The formula for inductive reactance is as follows:

$$X_L = 2\pi f L$$

Where:

X_L is inductive reactance in ohms.

2π is a constant in which the Greek letter π, called "pi" represents 3.1416 and $2 \times \pi =$ 6.28 approximately.

f   is frequency of the alternating current in Hz.

L   is inductance in henrys.

The following example problem illustrates the computation of X_L.

Given: f = 60 Hz
           L = 20 H
Solution: $X_L = 2\pi f L$
           $X_L = 6.28 \times 60$ Hz $\times 20$ H
           $X_L = 7,536$ Ω

Q3. What is the term for the opposition an inductor presents to ac?

Q4. What is the formula used to compute the value of this opposition?

Q5. What happens to the value of X_L as frequency increases?

Q6. What happens to the value of X_L as inductance decreases?

CAPACITORS AND ALTERNATING CURRENT

The four parts of figure 4-3 show the variation of the alternating voltage and current in a capacitive circuit, for each quarter of one cycle. The solid line represents the voltage across the capacitor, and the dotted line represents the current. The line running through the center is the zero, or reference point, for both the voltage and the current. The bottom line marks off the time of the cycle in terms of electrical degrees. Assume that the ac voltage has been acting on the capacitor for some time before the time represented by the starting point of the sine wave in the figure.

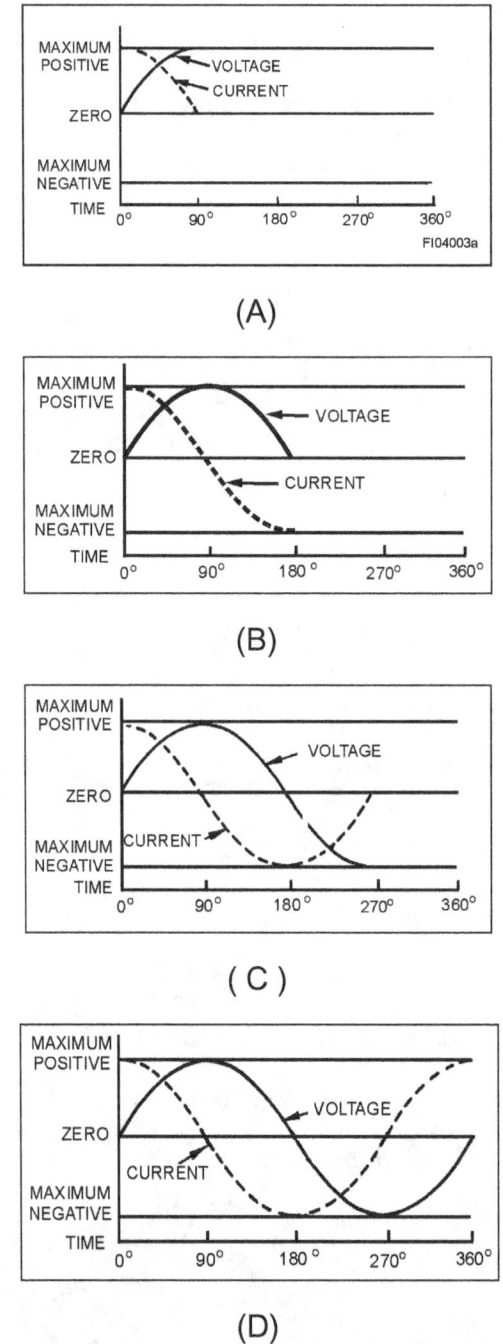

Figure 4-3.—Phase relationship of voltage and current in a capacitive circuit.

At the beginning of the first quarter-cycle (0° to 90°) the voltage has just passed through zero and is increasing in the positive direction. Since the zero point is the steepest part of the sine wave, the voltage is changing at its greatest rate. The charge on a capacitor varies directly with the voltage, and therefore the charge on the capacitor is also changing at its greatest rate at the beginning of the first quarter-cycle. In other words, the greatest number of electrons are moving off one plate and onto the other plate. Thus the capacitor current is at its maximum value, as part (A) of the figure shows.

As the voltage proceeds toward maximum at 90 degrees, its rate of change becomes less and less, hence the current must decrease toward zero. At 90 degrees the voltage across the capacitor is maximum, the capacitor is fully charged, and there is no further movement of electrons from plate to plate. That is why the current at 90 degrees is zero.

At the end of this first quarter-cycle the alternating voltage stops increasing in the positive direction and starts to decrease. It is still a positive voltage, but to the capacitor the decrease in voltage means that the plate which has just accumulated an excess of electrons must lose some electrons. The current flow, therefore, must reverse its direction. Part (B) of the figure shows the current curve to be below the zero line (negative current direction) during the second quarter-cycle (90° to 180°).

At 180 degrees the voltage has dropped to zero. This means that for a brief instant the electrons are equally distributed between the two plates; the current is maximum because the rate of change of voltage is maximum. Just after 180 degrees the voltage has reversed polarity and starts building up its maximum negative peak which is reached at the end of the third quarter-cycle (180° to 270°). During this third quarter-cycle the rate of voltage change gradually decreases as the charge builds to a maximum at 270 degrees. At this point the capacitor is fully charged and it carries the full impressed voltage. Because the capacitor is fully charged there is no further exchange of electrons; therefore, the current flow is zero at this point. The conditions are exactly the same as at the end of the first quarter-cycle (90°) but the polarity is reversed.

Just after 270 degrees the impressed voltage once again starts to decrease, and the capacitor must lose electrons from the negative plate. It must discharge, starting at a minimum rate of flow and rising to a maximum. This discharging action continues through the last quarter-cycle (270° to 360°) until the impressed-voltage has reached zero. At 360 degrees you are back at the beginning of the entire cycle, and everything starts over again.

If you examine the complete voltage and current curves in part D, you will see that the current always arrives at a certain point in the cycle 90 degrees ahead of the voltage, because of the charging and discharging action. You know that this time and place relationship between the current and voltage is called the phase relationship. The voltage-current phase relationship in a capacitive circuit is exactly opposite to that in an inductive circuit. <u>The current of a capacitor leads the voltage across the capacitor by 90 degrees</u>.

You realize that the current and voltage are both going through their individual cycles at the same time during the period the ac voltage is impressed. The current does not go through part of its cycle (charging or discharging), stop, and wait for the voltage to catch up. The amplitude and polarity of the voltage and the amplitude and direction of the current are continually changing. Their positions with respect to each other and to the zero line at any electrical instant-any degree between zero and 360 degrees-can be seen by reading upwards from the time-degree line. The current swing from the positive peak at zero degrees to the negative peak at 180 degrees is NOT a measure of the number of electrons, or the charge on the plates. It is a picture of the direction and strength of the current in relation to the polarity and strength of the voltage appearing across the plates.

At times it is convenient to use the word "ICE" to recall to mind the phase relationship of the current and voltage in capacitive circuits. I is the symbol for current, and in the word ICE it leads, or comes before, the symbol for voltage, E. C, of course, stands for capacitor. This memory aid is similar to the "ELI" used to remember the current and voltage relationship in an inductor. The phrase "ELI the ICE man" is helpful in remembering the phase relationship in both the inductor and capacitor.

Since the plates of the capacitor are changing polarity at the same rate as the ac voltage, the capacitor seems to pass an alternating current. Actually, the electrons do not pass through the dielectric, but their rushing back and forth from plate to plate causes a current flow in the circuit. It is convenient, however, to say that the alternating current flows "through" the capacitor. You know this is not true, but the expression avoids a lot of trouble when speaking of current flow in a circuit containing a capacitor. By the same short cut, you may say that the capacitor does not pass a direct current (if both plates are connected to a dc source, current will flow only long enough to charge the capacitor). With a capacitor type of hookup in a circuit containing both ac and dc, only the ac will be "passed" on to another circuit.

You have now learned two things to remember about a capacitor: <u>A capacitor will appear to conduct an alternating current and a capacitor will not conduct a direct current</u>.

Q7. *What effect does the capacitor have on a changing voltage?*

Q8. *What is the phase relationship between current and voltage in a capacitor?*

CAPACITIVE REACTANCE

So far you have been dealing with the capacitor as a device which passes ac and in which the only opposition to the alternating current has been the normal circuit resistance present in any conductor. However, capacitors themselves offer a very real opposition to current flow. This opposition arises from the fact that, at a given voltage and frequency, the number of electrons which go back and forth from plate to plate is limited by the storage ability-that is, the capacitance-of the capacitor. As the capacitance is increased, a greater number of electrons change plates every cycle, and (since current is a measure of the number of electrons passing a given point in a given time) the current is increased.

Increasing the frequency will also decrease the opposition offered by a capacitor. This occurs because the number of electrons which the capacitor is capable of handling at a given voltage will change plates more often. As a result, more electrons will pass a given point in a given time (greater current flow). The opposition which a capacitor offers to ac is therefore inversely proportional to frequency and to capacitance. This opposition is called CAPACITIVE REACTANCE. You may say that capacitive reactance decreases with increasing frequency or, for a given frequency, the capacitive reactance decreases with increasing capacitance. The symbol for capacitive reactance is X_C.

Now you can understand why it is said that the X_C varies inversely with the product of the frequency and capacitance. The formula is:

$$X_C = \frac{1}{2\pi f C}$$

Where:

X_C is capacitive reactance in ohms

f is frequency in Hertz

C is capacitance in farads

π is 6.28 (2 × 3.1416)

The following example problem illustrates the computation of X_C.

Given: f = 100 Hz
          C = 50 μF

Solution: $$X_C = \frac{1}{2\pi fC}$$

$$X_C = \frac{1}{6.28 \times 100 \text{ Hz} \times 50 \mu F}$$

$$X_C = \frac{1}{.0314} \Omega$$

$$X_C = 31.8 \Omega \text{ or } 32 \Omega$$

Q9. *What is the term for the opposition that a capacitor presents to ac?*

Q10. *What is the formula used to compute this opposition?*

Q11. *What happens to the value of X_C as frequency decreases?*

Q12. *What happens to the value of X_C as capacitance increases?*

REACTANCE, IMPEDANCE, AND POWER RELATIONSHIPS IN AC CIRCUITS

Up to this point inductance and capacitance have been explained individually in ac circuits. The rest of this chapter will concern the combination of inductance, capacitance, and resistance in ac circuits.

To explain the various properties that exist within ac circuits, the series RLC circuit will be used. Figure 4-4 is the schematic diagram of the series RLC circuit. The symbol shown in figure 4-4 that is marked E is the general symbol used to indicate an ac voltage source.

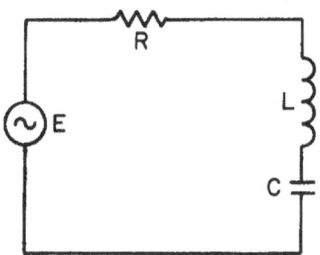

Figure 4-4.—Series RLC circuit.

REACTANCE

The effect of inductive reactance is to cause the current to lag the voltage, while that of capacitive reactance is to cause the current to lead the voltage. Therefore, since inductive reactance and capacitive reactance are exactly opposite in their effects, what will be the result when the two are combined? It is not hard to see that the net effect is a tendency to cancel each other, with the combined effect then equal to the difference between their values. This resultant is called REACTANCE; it is represented by the symbol X; and expressed by the equation $X = X_L - X_C$ or $X = X_C - X_L$. Thus, if a circuit contains 50 ohms of inductive reactance and 25 ohms of capacitive reactance in series, the net reactance, or X, is 50 ohms – 25 ohms, or 25 ohms of inductive reactance.

For a practical example, suppose you have a circuit containing an inductor of 100 µH in series with a capacitor of .001 µF, and operating at a frequency of 4 MHz. What is the value of net reactance, or X?

$$\text{Given:} \quad f = 4\,\text{MHz}$$
$$L = 100\,\mu H$$
$$C = .001\,\mu F$$

$$\text{Solution:} \quad X_L = 2\pi fL$$
$$X_L = 6.28 \times 4\,\text{MHz} \times 100\,\mu H$$
$$X_L = 2512\,\Omega$$
$$X_C = \frac{1}{2\pi fC}$$
$$X_C = \frac{1}{6.28 \times 4\,\text{MHz} \times .001\,\mu F}$$
$$X_C = \frac{1}{.02512}\,\Omega$$
$$X_C = 39.8\,\Omega$$
$$X = X_L - X_C$$
$$X = 2512\,\Omega - 39.8\,\Omega$$
$$X = 2472.2\,\Omega \ (\text{inductive})$$

Now assume you have a circuit containing a 100 - µH inductor in series with a .0002-µF capacitor, and operating at a frequency of 1 MHz. What is the value of the resultant reactance in this case?

Given: $f = 1\,\text{MHz}$
$L = 100\,\mu\text{H}$
$C = .0002\,\mu\text{F}$

Solution: $X_L = 2\pi f L$
$X_L = 6.28 \times 1\,\text{MHz} \times 100\,\mu\text{H}$
$X_L = 628\,\Omega$
$X_C = \dfrac{1}{2\pi f C}$
$X_C = \dfrac{1}{6.28 \times 1\,\text{MHz} \times .0002\,\mu\text{F}}$
$X_C = \dfrac{1}{.001256}\,\Omega$
$X_C = 796\,\Omega$
$X = X_C - X_L$
$X = 796\,\Omega - 628\,\Omega$
$X = 168\,\Omega\ \text{(capacitive)}$

You will notice that in this case the inductive reactance is smaller than the capacitive reactance and is therefore subtracted from the capacitive reactance.

These two examples serve to illustrate an important point: when capacitive and inductive reactance are combined in series, the smaller is always subtracted from the larger and the resultant reactance always takes the characteristics of the larger.

Q13. What is the formula for determining total reactance in a series circuit where the values of X_C and X_L are known?

Q14. What is the total amount of reactance (X) in a series circuit which contains an X_L of 20 ohms and an X_C of 50 ohms? (Indicate whether X is capacitive or inductive)

IMPEDANCE

From your study of inductance and capacitance you know how inductive reactance and capacitive reactance act to oppose the flow of current in an ac circuit. However, there is another factor, the resistance, which also opposes the flow of the current. Since in practice ac circuits containing reactance also contain resistance, the two combine to oppose the flow of current. This combined opposition by the resistance and the reactance is called the IMPEDANCE, and is represented by the symbol Z.

Since the values of resistance and reactance are both given in ohms, it might at first seem possible to determine the value of the impedance by simply adding them together. It cannot be done so easily, however. You know that in an ac circuit which contains only resistance, the current and the voltage will be in step (that is, in phase), and will reach their maximum values at the same instant. You also know that in an ac circuit containing only reactance the current will either lead or lag the voltage by one-quarter of a cycle or 90 degrees. Therefore, the voltage in a purely reactive circuit will differ in phase by 90 degrees from that in a purely resistive circuit and for this reason reactance and resistance are rot combined by simply adding them.

When reactance and resistance are combined, the value of the impedance will be greater than either. It is also true that the current will not be in step with the voltage nor will it differ in phase by exactly 90 degrees from the voltage, but it will be somewhere between the in-step and the 90-degree out-of-step conditions. The larger the reactance compared with the resistance, the more nearly the phase difference will approach 90°. The larger the resistance compared to the reactance, the more nearly the phase difference will approach zero degrees.

If the value of resistance and reactance cannot simply be added together to find the impedance, or Z, how is it determined? Because the current through a resistor is in step with the voltage across it and the current in a reactance differs by 90 degrees from the voltage across it, the two are at right angles to each other. They can therefore be combined by means of the same method used in the construction of a right-angle triangle.

Assume you want to find the impedance of a series combination of 8 ohms resistance and 5 ohms inductive reactance. Start by drawing a horizontal line, R, representing 8 ohms resistance, as the base of the triangle. Then, since the effect of the reactance is always at right angles, or 90 degrees, to that of the resistance, draw the line X_L, representing 5 ohms inductive reactance, as the altitude of the triangle. This is shown in figure 4-5. Now, complete the hypotenuse (longest side) of the triangle. Then, the hypotenuse represents the impedance of the circuit.

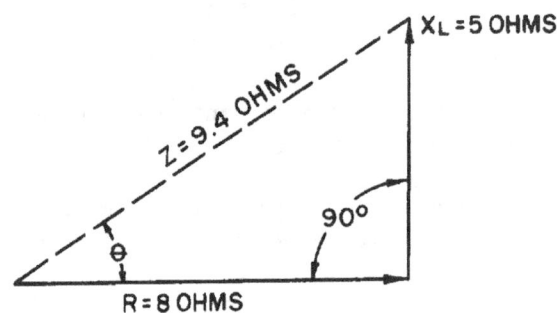

Figure 4-5.—Vector diagram showing relationship of resistance, inductive reactance, and impedance in a series circuit.

One of the properties of a right triangle is:

$$(\text{hypotenuse})^2 = (\text{base})^2 + (\text{altitude})^2$$

or,

$$\text{hypotenuse} = \sqrt{(\text{base})^2 + (\text{altitude})^2}$$

Applied to impedance, this becomes,

$$(\text{impedance})^2 = (\text{resistance})^2 + (\text{reactance})^2$$

or,

$$\text{impedance} = \sqrt{(\text{resistance})^2 + (\text{reactance})^2}$$

or,

$$Z = \sqrt{R^2 + X^2}$$

Now suppose you apply this equation to check your results in the example given above.

Given: $R = 8\Omega$
$X_L = 5\Omega$

Solution: $Z = \sqrt{R^2 + X_L^2}$
$Z = \sqrt{(8\Omega)^2 + (5\Omega)^2}$
$Z = \sqrt{64 + 25}\,\Omega$
$Z = \sqrt{89}\,\Omega$ (See the Appendix III for a square Root Table.)
$Z = 9.4\Omega$

When you have a capacitive reactance to deal with instead of inductive reactance as in the previous example, it is customary to draw the line representing the capacitive reactance in a downward direction. This is shown in figure 4-6. The line is drawn downward for capacitive reactance to indicate that it acts in a direction opposite to inductive reactance which is drawn upward. In a series circuit containing capacitive reactance the equation for finding the impedance becomes:

$$Z = \sqrt{R^2 + X_C^2}$$

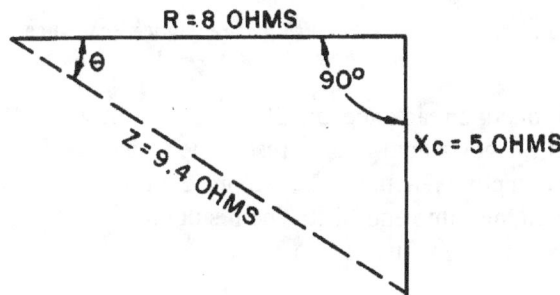

Figure 4-6.—Vector diagram showing relationship of resistance, capacitive reactance, and impedance in a series circuit.

In many series circuits you will find resistance combined with both inductive reactance and capacitive reactance. Since you know that the value of the reactance, X, is equal to the difference between the values of the inductive reactance, X_L, and the capacitive reactance, X_C, the equation for the impedance in a series circuit containing R, X_L, and X_C then becomes:

$$Z = \sqrt{R^2 + (X_L - X_C)^2}$$

or,

$$Z = \sqrt{R^2 + X^2}$$

(Note: The formulas $Z = \sqrt{R^2 + X_L^2}$, $Z = \sqrt{R^2 + X_C^2}$, and $Z = \sqrt{R^2 + X^2}$ can be used to calculate Z only if the resistance and reactance are connected in series.)

In figure 4-7 you will see the method which may be used to determine the impedance in a series circuit consisting of resistance, inductance, and capacitance.

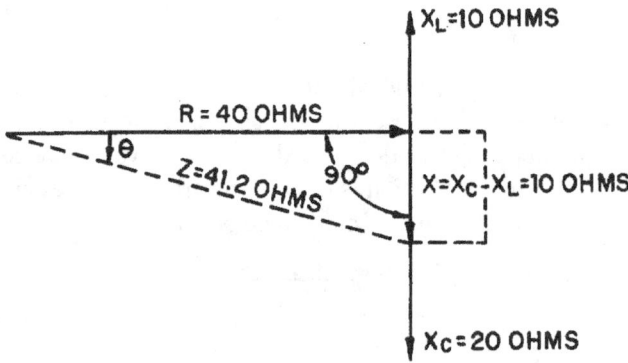

Figure 4-7.—Vector diagram showing relationship of resistance, reactance (capacitive and inductive), and impedance in a series circuit.

Assume that 10 ohms inductive reactance and 20 ohms capacitive reactance are connected in series with 40 ohms resistance. Let the horizontal line represent the resistance R. The line drawn upward from the end of R, represents the inductive reactance, X_L. Represent the capacitive reactance by a line drawn downward at right angles from the same end of R. The resultant of X_L and X_C is found by subtracting X_L from X_C. This resultant represents the value of X.

Thus:

$$X = X_C - X_L$$
$$X = 10 \text{ ohms}$$

The line, Z, will then represent the resultant of R and X. The value of Z can be calculated as follows:

Given: $X_L = 10 \ \Omega$
$X_C = 20 \ \Omega$
$R = 40 \ \Omega$

4-14

Solution:
$$X = X_c - X_L$$
$$X = 20\,\Omega - 10\,\Omega$$
$$X = 10\,\Omega$$
$$Z = \sqrt{R^2 + X^2}$$
$$Z = \sqrt{(40\,\Omega)^2 + (10\,\Omega^2)}$$
$$Z = \sqrt{1600 + 100}\,\Omega$$
$$Z = \sqrt{1700}\,\Omega$$
$$Z = 41.2\,\Omega$$

Q15. What term is given to total opposition to ac in a circuit?

Q16. What formula is used to calculate the amount of this opposition in a series circuit?

Q17. What is the value of Z in a series ac circuit where $X_L = 6$ ohms, $X_C = 3$ ohms, and $R = 4$ ohms?

OHMS LAW FOR AC

In general, Ohm's law cannot be applied to alternating-current circuits since it does not consider the reactance which is always present in such circuits. However, by a modification of Ohm's law which does take into consideration the effect of reactance we obtain a general law which is applicable to ac circuits. Because the impedance, Z, represents the combined opposition of all the reactances and resistances, this general law for ac is,

$$I = \frac{E}{Z}$$

this general modification applies to alternating current flowing in any circuit, and any one of the values may be found from the equation if the others are known.

For example, suppose a series circuit contains an inductor having 5 ohms resistance and 25 ohms inductive reactance in series with a capacitor having 15 ohms capacitive reactance. If the voltage is 50 volts, what is the current? This circuit can be drawn as shown in figure 4-8.

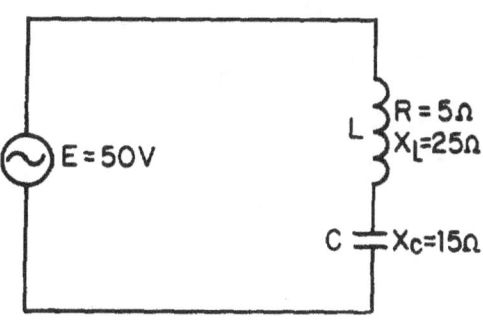

Figure 4-8.—Series LC circuit.

4-15

Given: $R = 5\,\Omega$
$X_L = 25\,\Omega$
$X_C = 15\,\Omega$
$E = 50\,V$

Solution: $X = X_L - X_C$
$X = 25\,\Omega - 15\,\Omega$
$X = 10\,\Omega$
$Z = \sqrt{R^2 + X^2}$
$Z = \sqrt{(5\,\Omega)^2 + (10\,\Omega)^2}$
$Z = \sqrt{25 + 100}\,\Omega$
$Z = \sqrt{125}\,\Omega$
$Z = 11.2\,\Omega$
$I = \dfrac{E}{Z}$
$I = \dfrac{50\,V}{11.2\,\Omega}$
$I = 4.46\,A$

Now suppose the circuit contains an inductor having 5 ohms resistance and 15 ohms inductive reactance in series with a capacitor having 10 ohms capacitive reactance. If the current is 5 amperes, what is the voltage?

Given: $R = 5\,\Omega$
$X_L = 15\,\Omega$
$X_C = 10\,\Omega$
$I = 5\,A$

Solution: $X = X_L - X_C$
$X = 15\,\Omega - 10\,\Omega$
$X = 5\,\Omega$
$Z = \sqrt{R^2 + X^2}$
$Z = \sqrt{(5\,\Omega)^2 + (5\,\Omega)^2}$
$Z = \sqrt{25 + 25}\,\Omega$
$Z = \sqrt{50}\,\Omega$
$Z = 7.07\,\Omega$
$E = IZ$
$E = 5\,A \times 7.07\,\Omega$
$E = 35.35\,V$

Q18. What are the Ohm's law formulas used in an ac circuit to determine voltage and current?

POWER IN AC CIRCUITS

You know that in a direct current circuit the power is equal to the voltage times the current, or $P = E \times I$. If a voltage of 100 volts applied to a circuit produces a current of 10 amperes, the power is 1000 watts. This is also true in an ac circuit when the current and voltage are in phase; that is, when the circuit is effectively resistive. But, if the ac circuit contains reactance, the current will lead or lag the voltage by a certain amount (the phase angle). When the current is out of phase with the voltage, the power indicated by the product of the applied voltage and the total current gives only what is known as the APPARENT POWER. The TRUE POWER depends upon the phase angle between the current and voltage. The symbol for phase angle is θ (Theta).

When an alternating voltage is impressed across a capacitor, power is taken from the source and stored in the capacitor as the voltage increases from zero to its maximum value. Then, as the impressed voltage decreases from its maximum value to zero, the capacitor discharges and returns the power to the source. Likewise, as the current through an inductor increases from its zero value to its maximum value the field around the inductor builds up to a maximum, and when the current decreases from maximum to zero the field collapses and returns the power to the source. You can see therefore that no power is used up in either case, since the power alternately flows to and from the source. This power that is returned to the source by the reactive components in the circuit is called REACTIVE POWER.

In a purely resistive circuit <u>all of the power is consumed and none is returned to the source</u>; in a purely reactive circuit <u>no power is consumed and all of the power is returned to the source</u>. It follows that in a circuit which contains both resistance and reactance there must be some power dissipated in the resistance as well as some returned to the source by the reactance. In figure 4-9 you can see the relationship between the voltage, the current, and the power in such a circuit. The part of the power curve which is shown below the horizontal reference line is the result of multiplying a positive instantaneous value of current by a negative instantaneous value of the voltage, or vice versa. As you know, the product obtained by multiplying a positive value by a negative value will be negative. Therefore the power at that instant must be considered as negative power. In other words, during this time the reactance was returning power to the source.

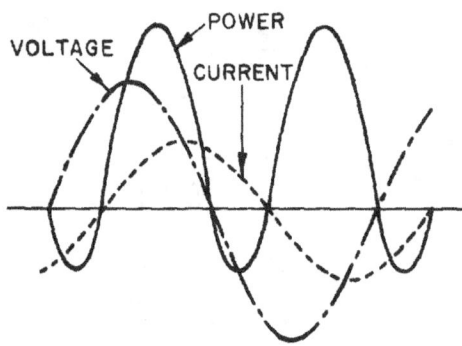

Figure 4-9.—Instantaneous power when current and voltage are out of phase.

The instantaneous power in the circuit is equal to the product of the applied voltage and current through the circuit. When the voltage and current are of the same polarity they are acting together and taking power from the source. When the polarities are unlike they are acting in opposition and power is being returned to the source. Briefly then, in an ac circuit which contains reactance as well as resistance, the apparent power is reduced by the power returned to the source, so that in such a circuit the net power, or <u>true power</u>, is always less than the apparent power.

Calculating True Power in AC Circuits

As mentioned before, the true power of a circuit is the power actually used in the circuit. This power, measured in watts, is the power associated with the total resistance in the circuit. To calculate true power, the voltage and current associated with the resistance must be used. Since the voltage drop across the resistance is equal to the resistance multiplied by the current through the resistance, true power can be calculated by the formula:

$$\text{True Power} = (I_R)^2 R$$

Where: True Power is measured in watts
I_R is resistive current in amperes
R is resistance in ohms

For example, find the true power of the circuit shown in figure 4-10.

4-18

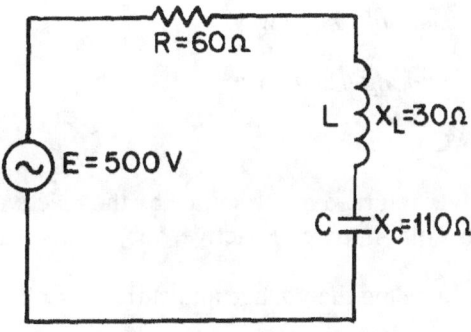

Figure 4-10.—Example circuit for determining power.

Given: $R = 60\,\Omega$
$X_L = 30\,\Omega$
$X_C = 110\,\Omega$
$E = 500\,V$

Solution: $X = X_C - X_L$
$X = 110\,\Omega - 30\,\Omega$
$X = 80\,\Omega$
$Z = \sqrt{R^2 + X^2}$
$Z = \sqrt{(60\,\Omega)^2 + (80\,\Omega)^2}$
$Z = \sqrt{3600 + 6400}\,\Omega$
$Z = \sqrt{10{,}000}\,\Omega$
$Z = 100\,\Omega$
$I = \dfrac{E}{Z}$
$I = \dfrac{500\,V}{100\,\Omega}$
$I = 5\,A$

Since the current in a series circuit is the same in all parts of the circuit:

$$\text{True Power} = (I_R)^2 R$$
$$\text{True Power} = (5\,A)^2 \times 60\,\Omega$$
$$\text{True Power} = 1500\,\text{watts}$$

Q19. What is the true power in an ac circuit?

Q20. What is the unit of measurement of true power?

Q21. What is the formula for calculating true power?

Calculating Reactive Power in AC Circuits

The reactive power is the power returned to the source by the reactive components of the circuit. This type of power is measured in Volt-Amperes-Reactive, abbreviated var.

Reactive power is calculated by using the voltage and current associated with the circuit reactance.

Since the voltage of the reactance is equal to the reactance multiplied by the reactive current, reactive power can be calculated by the formula:

$$\text{Reactive Power} = (I_X)^2 X$$

Where: Reactive power is measured in volt-amperes-reactive.

I_X is reactive current in amperes.

X is total reactance in ohms.

Another way to calculate reactive power is to calculate the inductive power and capacitive power and subtract the smaller from the larger.

$$\text{Reactive Power} = (I_L)^2 X_L - (I_C)^2 X_C$$
$$\text{or}$$
$$(I_C)^2 X_C - (I_L)^2 X_L$$

Where: Reactive power is measured in volt-amperes-reactive.

I_C is capacitive current in amperes.

X_C is capacitive reactance in ohms.

I_L is inductive current in amperes.

X_L is inductive reactance in ohms.

Either one of these formulas will work. The formula you use depends upon the values you are given in a circuit.

For example, find the reactive power of the circuit shown in figure 4-10.

$$\text{Given:} \quad X_L = 30\,\Omega$$
$$X_C = 110\,\Omega$$
$$X = 80\,\Omega$$
$$I = 5\,A$$

Since this is a series circuit, current (I) is the same in all parts of the circuit.

$$\text{Solution:} \quad \text{Reactive power} = (I_X)^2 X$$
$$\text{Reactive power} = (5A)^2 \times 80\,\Omega$$
$$\text{Reactive power} = 2{,}000\,\text{var}$$

To prove the second formula also works,
$$\text{Reactive power} = (I_C)^2 X_C - (I_L)^2 X_L$$
$$\text{Reactive power} = (5A)^2 \times 110\,\Omega - (5A)^2 \times 30\,\Omega$$
$$\text{Reactive power} = 2{,}750\,\text{var} - 750\,\text{var}$$
$$\text{Reactive power} = 2000\,\text{var}$$

Q22. What is the reactive power in an ac circuit?

Q23. What is the unit of measurement for reactive power?

Q24. What is the formula for computing reactive power?

Calculating Apparent Power in AC Circuits.

Apparent power is the power that appears to the source because of the circuit impedance. Since the impedance is the total opposition to ac, the apparent power is that power the voltage source "sees." Apparent power is the combination of true power and reactive power. Apparent power is not found by simply adding true power and reactive power just as impedance is not found by adding resistance and reactance.

To calculate apparent power, you may use either of the following formulas:

$$\text{Apparent power} = (I_Z)^2 Z$$

Where: Apparent power is measured in VA (volt-amperes)

I_Z is impedance current in amperes.

Z is impedance in ohms.

or

$$\text{Apparent power} = \sqrt{(\text{True power})^2 + (\text{reactive power})^2}$$

For example, find the apparent power for the circuit shown in figure 4-10.

Given: $Z = 100\ \Omega$
$I = 5\ A$

Recall that current in a series circuit is the same in all parts of the circuit.

Solution:

$\text{Apparent Power} = (I_Z)^2 Z$

$\text{Apparent power} = (5\ A)^2 \times 100\ \Omega$

$\text{Apparent power} = 2500\ VA$

or

Given:
True power = 1500 W
Reactive power = 2000 var

$\text{Apparent power} = \sqrt{(\text{True power})^2 + (\text{reactive power})^2}$

$\text{Apparent power} = \sqrt{(1500\ W)^2 + (2000\ var)^2}$

$\text{Apparent power} = \sqrt{625 \times 10^4}\ VA$

$\text{Apparent power} = 2500\ VA$

Q25. What is apparent power?

Q26. What is the unit of measurement for apparent power?

Q27. What is the formula for apparent power?

Power Factor

The POWER FACTOR is a number (represented as a decimal or a percentage) that represents the portion of the apparent power dissipated in a circuit.

If you are familiar with trigonometry, the easiest way to find the power factor is to find the cosine of the phase angle (θ). The cosine of the phase angle is equal to the power factor.

You do not need to use trigonometry to find the power factor. Since the power dissipated in a circuit is true power, then:

$$\text{Apparent Power} \times PF = \text{True Power},$$

Therefore, $$PF = \frac{\text{True Power}}{\text{Apparent Power}}$$

If true power and apparent power are known you can use the formula shown above.

Going one step further, another formula for power factor can be developed. By substituting the equations for true power and apparent power in the formula for power factor, you get:

$$PF = \frac{(I_R)^2 R}{(I_Z)^2 Z}$$

Since current in a series circuit is the same in all parts of the circuit, I_R equals I_Z. Therefore, in a series circuit,

$$PF = \frac{R}{Z}$$

For example, to compute the power factor for the series circuit shown in figure 4-10, any of the above methods may be used.

Given:

True Power = 1500 V
Apparent Power = 2500 VA

Solution:
$$PF = \frac{\text{True Power}}{\text{Apparent Power}}$$
$$PF = \frac{1500 \text{ W}}{2500 \text{ VA}}$$
$$PF = .6$$

Another method:

Given: $R = 60\ \Omega$
          $Z = 100\ \Omega$

Solution: $PF = \dfrac{R}{Z}$

          $PF = \dfrac{60\ \Omega}{100\ \Omega}$

          $PF = .6$

If you are familiar with trigonometry you can use it to solve for angle θ and the power factor by referring to the tables in appendices V and VI.

Given: $R = 60\ \Omega$
          $X = 80\ \Omega$

Solution: $\tan \theta = \dfrac{X}{R}$

          $\tan \theta = \dfrac{80\ \Omega}{60\ \Omega}$

          $\tan \theta = 1.333$

          $\theta = 53.1°$

          $PF = \cos \theta$

          $PF = .6$

NOTE: As stated earlier the power factor can be expressed as a decimal or percentage. In this example the decimal number .6 could also be expressed as 60%.

Q28. What is the power factor of a circuit?

Q29. What is a general formula used to calculate the power factor of a circuit?

Power Factor Correction

The apparent power in an ac circuit has been described as the power the source "sees". As far as the source is concerned the apparent power is the power that must be provided to the circuit. You also know that the true power is the power actually used in the circuit. The difference between apparent power and true power is wasted because, in reality, only true power is consumed. The ideal situation would be for apparent power and true power to be equal. If this were the case the power factor would be 1 (unity) or 100 percent. There are two ways in which this condition can exist. (1) If the circuit is purely resistive or (2) if the circuit "appears" purely resistive to the source. To make the circuit appear purely resistive there must be no reactance. To have no reactance in the circuit, the inductive reactance (X_L) and capacitive reactance (X_C) must be equal.

Remember: $X = X_L - X_C$

Therefore, when

$X_L = X_C$, $X = 0$

The expression "correcting the power factor" refers to reducing the reactance in a circuit.

The ideal situation is to have no reactance in the circuit. This is accomplished by adding capacitive reactance to a circuit which is inductive and inductive reactance to a circuit which is capacitive. For example, the circuit shown in figure 4-10 has a total reactance of 80 ohms capacitive and the power factor was .6 or 60 percent. If 80 ohms of inductive reactance were added to this circuit (by adding another inductor) the circuit would have a total reactance of zero ohms and a power factor of 1 or 100 percent. The apparent and true power of this circuit would then be equal.

Q30. An ac circuit has a total reactance of 10 ohms inductive and a total resistance of 20 ohms. The power factor is .89. What would be necessary to correct the power factor to unity?

SERIES RLC CIRCUITS

The principles and formulas that have been presented in this chapter are used in all ac circuits. The examples given have been series circuits.

This section of the chapter will not present any new material, but will be an example of using all the principles presented so far. You should follow each example problem step by step to see how each formula used depends upon the information determined in earlier steps. When an example calls for solving for square root, you can practice using the square-root table by looking up the values given.

The example series RLC circuit shown in figure 4-11 will be used to solve for X_L, X_C, X, Z, I_T, true power, reactive power, apparent power, and power factor.

The values solved for will be rounded off to the nearest whole number.

First solve for X_L and X_C.

Given: $f = 60\,\text{Hz}$
         $L = 27\,\text{mH}$
         $C = 380\,\mu\text{F}$

Solution: $X_L = 2\pi fL$
            $X_L = 6.28 \times 60\,\text{Hz} \times 27\,\text{mH}$
            $X_L = 10\,\Omega$
            $X_C = \dfrac{1}{2\pi fc}$
            $X_C = \dfrac{1}{6.28 \times 60\,\text{Hz} \times 380\,\mu\text{F}}$
            $X_C = \dfrac{1}{0.143}\,\Omega$
            $X_C = 7\,\Omega$

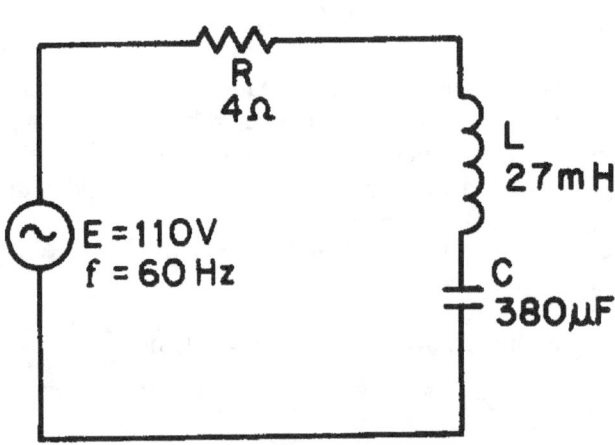

Figure 4-11.—Example series RLC circuit

Now solve for X

Given: $X_C = 7\,\Omega$
         $X_L = 10\,\Omega$

Solution: $X = X_L - X_C$
            $X = 10\,\Omega - 7\,\Omega$
            $X = 3\,\Omega$ (Inductive)

Use the value of X to solve for Z.

$$\text{Given:} \quad X = 3\,\Omega$$
$$R = 4\,\Omega$$

$$\text{Solution:} \quad Z = \sqrt{X^2 + R^2}$$
$$Z = \sqrt{(3\,\Omega)^2 + (4\,\Omega)^2}$$
$$Z = \sqrt{9 + 16}\,\Omega$$
$$Z = \sqrt{25}\,\Omega$$
$$Z = 5\,\Omega$$

This value of Z can be used to solve for total current (I_T).

$$\text{Given:} \quad Z = 5\,\Omega$$
$$E = 110\,V$$

$$\text{Solution:} \quad I_T = \frac{E}{Z}$$
$$I_T = \frac{110\,V}{5\,\Omega}$$
$$I_T = 22\,A$$

Since current is equal in all parts of a series circuit, the value of I_T can be used to solve for the various values of power.

Given:
$$I_T = 22\text{ A}$$
$$R = 4\Omega$$
$$X = 3\Omega$$
$$Z = 5\Omega$$

Solution:
$$\text{True Power} = (I_R)^2 R$$
$$\text{True Power} = (22\text{ A})^2 \times 4\Omega$$
$$\text{True Power} = 1936\text{ W}$$

$$\text{Reactive power} = (I_X)^2 X$$
$$\text{Reactive power} = (22\text{ A})^2 \times 3\Omega$$
$$\text{Reactive power} = 1452\text{ var}$$

$$\text{Apparent power} = (I_Z)^2 Z$$
$$\text{Apparent Power} = (22\text{ A})^2 \times 5\Omega$$
$$\text{Apparent Power} = 2420\text{ VA}$$

The power factor can now be found using either apparent power and true power or resistance and impedance. The mathematics in this example is easier if you use impedance and resistance.

Given:
$$R = 4\Omega$$
$$Z = 5\Omega$$

Solution:
$$PF = \frac{R}{Z}$$
$$PF = \frac{4\Omega}{5\Omega}$$
$$PF = .8 \text{ or } 80\%$$

PARALLEL RLC CIRCUITS

When dealing with a parallel ac circuit, you will find that the concepts presented in this chapter for series ac circuits still apply. There is one major difference between a series circuit and a parallel circuit that must be considered. The difference is that current is the same in all parts of a series circuit, whereas voltage is the same across all branches of a parallel circuit. Because of this difference, the total impedance of a parallel circuit must be computed on the basis of the current in the circuit.

You should remember that in the series RLC circuit the following three formulas were used to find reactance, impedance, and power factor:

$$X = X_L - X_C \text{ or } X = X_C - X_L$$
$$Z = \sqrt{(I_R)^2 + X^2}$$
$$PF = \frac{R}{Z}$$

When working with a parallel circuit you must use the following formulas instead:

$$I_X = I_L - I_C \text{ or } I_X = I_C - I_L$$
$$I_Z = \sqrt{(I_R)^2 + (I_X)^2}$$
$$PF = \frac{I_R}{I_Z}$$

(The impedance of a parallel circuit is found by the formula $Z = \frac{E}{I_Z}$)

NOTE: If no value for E is given in a circuit, any value of E can be assumed to find the values of I_L, I_C, I_X, I_R, and I_Z. The same value of voltage is then used to find impedance.

For example, find the value of Z in the circuit shown in figure 4-12.

Given: $E = 300$ V
$R = 100 \ \Omega$
$X_L = 50 \ \Omega$
$X_C = 150 \ \Omega$

The first step in solving for Z is to calculate the individual branch currents.

4-29

Solution: $I_R = \dfrac{E}{R}$

$I_R = \dfrac{300\text{ V}}{100\ \Omega}$

$I_R = 3\text{ A}$

$I_L = \dfrac{E}{X_L}$

$I_L = \dfrac{300\text{ V}}{50\ \Omega}$

$I_L = 6\text{ A}$

$I_C = \dfrac{E}{X_C}$

$I_C = \dfrac{300\text{ V}}{150\ \Omega}$

$I_C = 2\text{ A}$

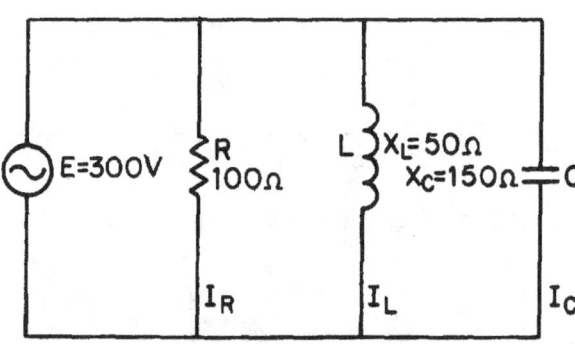

Figure 4-12.—Parallel RLC circuit.

Using the values for I_R, I_L, and I_C, solve for I_X and I_Z.

$I_X = I_L - I_C$

$I_X = 6\text{ A} - 2\text{ A}$

$I_X = 4\text{ A (inductive)}$

$I_Z = \sqrt{(I_R)^2 + (I_X)^2}$

$I_Z = \sqrt{(3\text{ A})^2 + (4\text{ A})^2}$

$I_Z = \sqrt{25}\text{ A}$

$I_Z = 5\text{ A}$

Using this value of I_Z, solve for Z.

$$Z = \frac{E}{I_Z}$$

$$Z = \frac{300 \text{ V}}{5 \text{ A}}$$

$$Z = 60 \, \Omega$$

If the value for E were not given and you were asked to solve for Z, any value of E could be assumed. If, in the example problem above, you assume a value of 50 volts for E, the solution would be:

Given: $R = 100 \, \Omega$
$X_L = 50 \, \Omega$
$X_C = 150 \, \Omega$
$E = 50 \text{ V (assumed)}$

First solve for the values of current in the same manner as before.

Solution: $I_R = \frac{E}{R}$

$$I_R = \frac{50 \text{ V}}{100 \, \Omega}$$

$$I_R = .5 \text{ A}$$

$$I_L = \frac{E}{X_L}$$

$$I_L = \frac{50 \text{ V}}{50 \, \Omega}$$

$$I_L = 1 \text{ A}$$

$$I_C = \frac{E}{X_C}$$

$$I_C = \frac{50 \text{ V}}{150 \, \Omega}$$

$$I_C = .33 \text{ A}$$

Solve for I_X and I_Z.

$$I_X = I_L - I_C$$
$$I_X = 1A - .33A$$
$$I_X = .67A \text{ (Inductive)}$$
$$I_Z = \sqrt{(I_R)^2 + (I_X)^2}$$
$$I_Z = \sqrt{(0.5A)^2 + (0.67A)^2}$$
$$I_Z = \sqrt{0.6989A}$$
$$I_Z = 0.836A$$

Solve for Z.

$$Z = \frac{E}{I_Z}$$
$$Z = \frac{50V}{.836A}$$
$$Z = 60\,\Omega \text{ (rounded off)}$$

When the voltage is given, you can use the values of currents, I_R, I_X, and I_Z, to calculate for the true power, reactive power, apparent power, and power factor. For the circuit shown in figure 4-12, the calculations would be as follows.

To find true power,

Given: $R = 100\,\Omega$
$I_R = 3A$

Solution:
$$\text{True Power} = (I_R)^2 X$$
$$\text{True Power} = (3A)^2 \times 75\,\Omega$$
$$\text{True Power} = 900\,W$$

To find reactive power, first find the value of reactance (X).

Given: $E = 300\,V$

$I_X = 4\,A$ (Inductive)

Solution: $X = \dfrac{E}{I_X}$

$X = \dfrac{300\,V}{4\,A}$

$X = 75\,\Omega$ (Inductive)

Reactive power $= (I_X)^2 X$

Reactive power $= (4\,A)^2 \times 75\,\Omega$

Reactive power $= 1200$ var

To find apparent power,

Given: $Z = 60\,\Omega$

$I_Z = 5\,A$

Solution:

Apparent Power $= (I_Z)^2 Z$

Apparent Power $= (5\,A)^2 \times 60\,\Omega$

Apparent Power $= 1500$ VA

The power factor in a parallel circuit is found by either of the following methods.

Given:

True Power = 900 W

Apparent Power = 1500 VA

Solution:
$$PF = \frac{\text{true power}}{\text{apparent power}}$$
$$PF = \frac{900 \text{ W}}{1500 \text{ VA}}$$
$$PF = .6$$

or

Given:
$$I_R = 3 \text{ A}$$
$$I_Z = 5 \text{ A}$$

Solution:
$$PF = \frac{I_R}{I_Z}$$
$$PF = \frac{3 \text{ A}}{5 \text{ A}}$$
$$PF = .6$$

Q31. *What is the difference between calculating impedance in a series ac circuit and in a parallel ac circuit?*

SUMMARY

With the completion of this chapter you now have all the building blocks for electrical circuits. The subjects covered from this point on will be based upon the concepts and relationships that you have learned. The following summary is a brief review of the subjects covered in this chapter.

INDUCTANCE IN AC CIRCUITS—An inductor in an ac circuit opposes any change in current flow just as it does in a dc circuit.

PHASE RELATIONSHIPS OF AN INDUCTOR—The current lags the voltage by 90° in an inductor (ELI).

INDUCTIVE REACTANCE—The opposition an inductor offers to ac is called inductive reactance. It will increase if there is an increase in frequency or an increase in inductance. The symbol is X_L, and the formula is $X_L = 2\pi fL$.

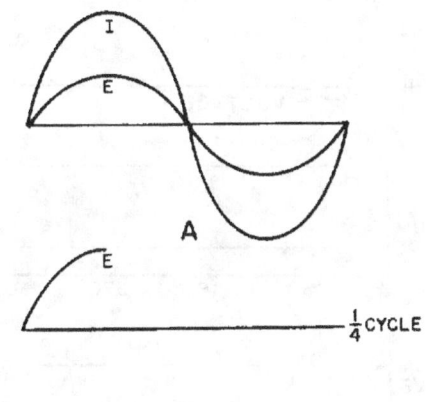

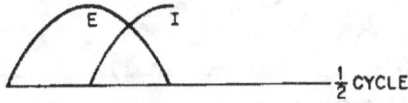

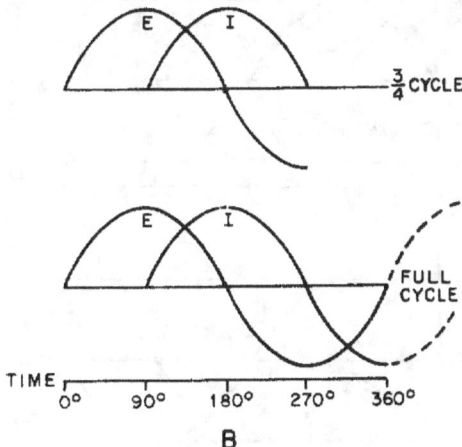

CAPACITANCE IN AC CIRCUITS—A capacitor in an ac circuit opposes any change in voltage just as it does in a dc circuit.

PHASE RELATIONSHIPS OF A CAPACITOR—The current leads the voltage by 90° in a capacitor (ICE).

CAPACITIVE REACTANCE—The opposition a capacitor offers to ac is called capacitive reactance. Capacitive reactance will decrease if there is an increase in frequency or an increase in capacitance. The symbol is X_C and the formula is

$$X_C = \frac{1}{2\pi f C}$$

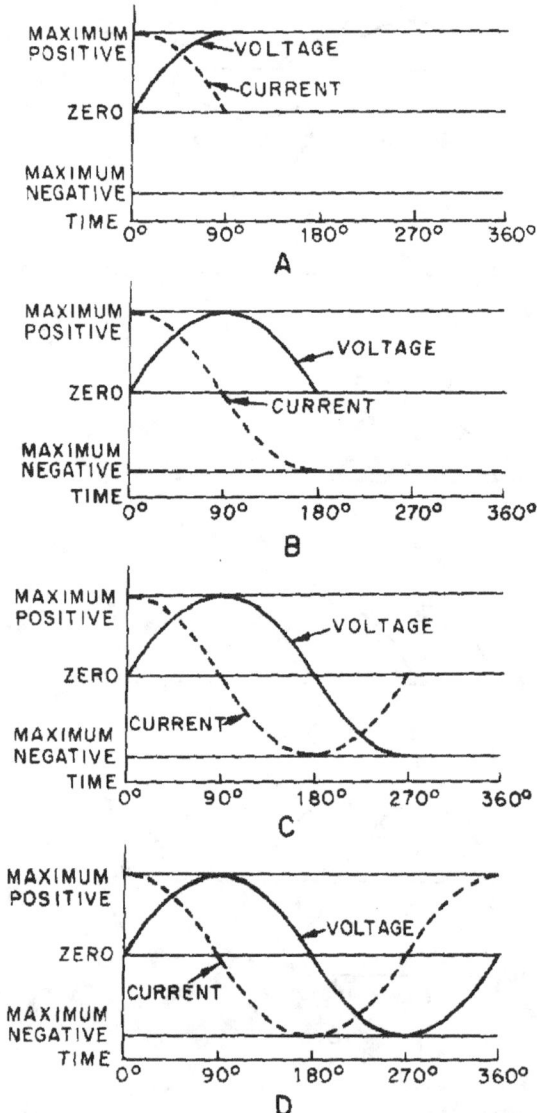

TOTAL REACTANCE—The total reactance of a series ac circuit is determined by the formula $X = X_L - X_C$ or $X = X_C - X_L$. The total reactance in a series circuit is either capacitive or inductive depending upon the largest value of X_C and X_L. In a parallel circuit the reactance is determined by

$$\frac{E}{I_X},$$

where $I_X = I_C - I_L$ or $I_X = I_L - I_C$. The reactance in a parallel circuit is either capacitive or inductive depending upon the largest value of I_L and I_C.

IMPEDANCE – The total opposition to a.c. is called impedance. The symbol is Z. In a series circuit $Z = \sqrt{R^2 + X^2}$. In a parallel circuit $I_Z = \sqrt{(I_R)^2 + (I_X)^2}$ and $Z = \dfrac{E}{I_Z}$.

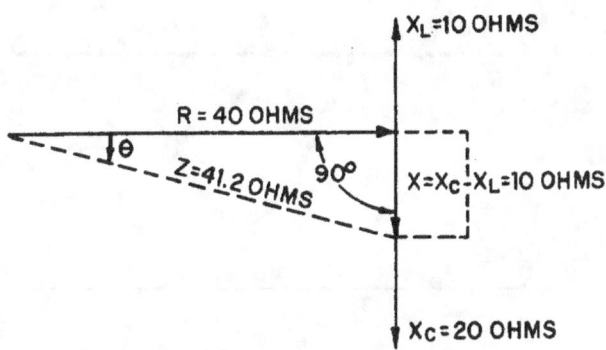

PHASE ANGLE—The number of degrees that current leads or lags voltage in an ac circuit is called the phase angle. The symbol is θ.

OHM'S LAW FORMULAS FOR AC—The formulas derived for Ohm's law used in ac are: $E = IZ$ and $I = E/Z$.

TRUE POWER—The power dissipated across the resistance in an ac circuit is called true power. It is measured in watts and the formula is: True Power = $(I_R)^2 R$.

REACTIVE POWER—The power returned to the source by the reactive elements of the circuit is called reactive power. It is measured in volt-amperes reactive (var). The formula is: Reactive Power = $(I_X)^2 X$.

APPARENT POWER—The power that appears to the source because of circuit impedance is called apparent power. It is the combination of true power and reactive power and is measured in volt-amperes (VA). The formulas are:

$$\text{Apparent Power} = (I_Z)^2 Z$$
$$\text{Apparent Power} = \sqrt{(\text{true power})^2 + (\text{reactive power})^2}$$

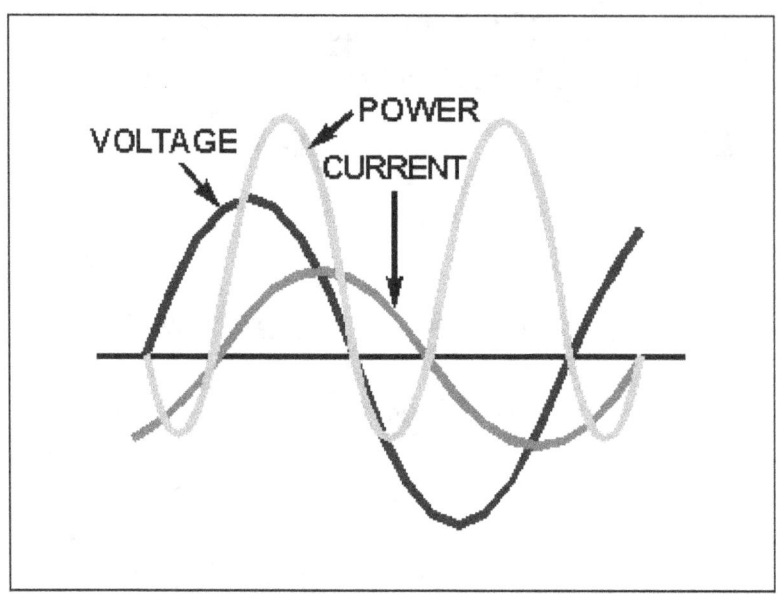

POWER FACTOR—The portion of the apparent power dissipated in a circuit is called the power factor of the circuit. It can be expressed as a decimal or a percentage. The formulas for power

$$\text{factor are } PF = \frac{\text{true power}}{\text{apparent power}} \text{ or } PF = \cos\theta. \text{ In a series circuit, } PF = \frac{R}{Z}. \text{ In a parallel circuit, } Pf = \frac{I_R}{I_Z}.$$

POWER FACTOR CORRECTION—To reduce losses in a circuit the power factor should be as close to unity or 100% as possible. This is done by adding capacitive reactance to a circuit when the total reactance is inductive. If the total reactance is capacitive, inductive reactance is added in the circuit.

ANSWERS TO QUESTIONS Q1. THROUGH Q31.

A1. *An inductor opposes a change in current.*

A2. *Current lags voltage by 90° (ELI).*

A3. *Inductive reactance.*

A4. *$X_L = 2\pi f L$.*

A5. *X_L increases.*

A6. *X_L decreases.*

A7. The capacitor opposes any change in voltage.

A8. Current leads voltage by 90° (ICE).

A9. Capacitive reactance.

A10.
$$X = \frac{1}{2\pi fC}$$

A11. X_C increases.

A12. X_C decreases.

A13. $X = X_L - X_C$ or $X = X_C - X_L$

A14. 30 Ω (capacitive).

A15. Impedance.

A16.
$$Z = \sqrt{R^2 + X^2}.$$

A17. $Z = 5\Omega$.

A18.
$$E = IZ$$
$$I = \frac{E}{Z}.$$

A19. True power is the power dissipated in the resistance of the circuit or the power actually used in the circuit.

A20. Watt.

A21. True Power = $(I_R)^2 R$.

A22. Reactive power is the power returned to the source by the reactive components of the circuit.

A23. var.

A24.
$$\text{Reactive Power} = (I_X)^2 X \text{ or}$$
$$(I_C)^2 X_C - (I_L)^2 X_L \text{ or}$$
$$(I_L)^2 X_L - (I_C)^2 X_C.$$

A25. The power that appears to the source because of circuit impedance, or the combination of true power and reactive power.

A26. VA (volt-amperes).

A27.
$$\text{Apparent power} = (I_Z)^2 Z \text{ or}$$
$$\sqrt{(\text{true power})^2 + (\text{reactive power})^2}$$

A28. PF is a number representing the portion of apparent power actually dissipated in a circuit.

A29.
$$PF = \frac{\text{true power}}{\text{apparent power}} \text{ or } PF = \cos\theta.$$

A30. Add 10 ohms of capacitive reactance to the circuit.

A31. In a series circuit impedance is calculated from the values of resistance and reactance. In a parallel circuit, the values of resistive current and reactive current must be used to calculate total current (impedance current) and this value must be divided into the source voltage to calculate the impedance.

CHAPTER 5
TRANSFORMERS

LEARNING OBJECTIVES

Upon completion of this chapter you will be able to:

1. State the meaning of "transformer action."

2. State the physical characteristics of a transformer, including the basic parts, common core materials, and main core types.

3. State the names given to the source and load windings of a transformer.

4. State the difference in construction between a high- and a low-voltage transformer.

5. Identify transformer symbols as to the type of transformer each symbol represents and the method used to denote transformer phasing.

6. State the meaning of a "no-load condition" and "exciting current" relative to a transformer.

7. State what causes voltage to be developed across the secondary of a transformer and the effect of cemf in a transformer.

8. State the meaning of leakage flux and its effect on the coefficient of coupling.

9. Identify a transformer as step up or step down and state the current ratio of a transformer when given the turns ratio.

10. Solve for primary voltage, secondary voltage, primary current and number of turns in the secondary given various transformer values.

11. State the mathematical relationship between the power in the primary and the power in the secondary of a transformer and compute efficiency of a transformer.

12. State the three power losses in a transformer.

13. State the reason a transformer should not be operated at a lower frequency than that specified for the transformer.

14. List five different types of transformers according to their applications.

15. State the standard color coding for a power transformer.

16. State the general safety precautions you should observe when working with transformers and other electrical components.

TRANSFORMERS

The information in this chapter is on the construction, theory, operation, and the various uses of transformers. Safety precautions to be observed by a person working with transformers are also discussed.

A TRANSFORMER is a device that transfers electrical energy from one circuit to another by electromagnetic induction (transformer action). The electrical energy is always transferred without a change in frequency, but may involve changes in magnitudes of voltage and current. Because a transformer works on the principle of electromagnetic induction, it must be used with an input source voltage that varies in amplitude. There are many types of power that fit this description; for ease of explanation and understanding, transformer action will be explained using an ac voltage as the input source.

In a preceding chapter you learned that alternating current has certain advantages over direct current. One important advantage is that when ac is used, the voltage and current levels can be increased or decreased by means of a transformer.

As you know, the amount of power used by the load of an electrical circuit is equal to the current in the load times the voltage across the load, or $P = EI$. If, for example, the load in an electrical circuit requires an input of 2 amperes at 10 volts (20 watts) and the source is capable of delivering only 1 ampere at 20 volts, the circuit could not normally be used with this particular source. However, if a transformer is connected between the source and the load, the voltage can be decreased (stepped down) to 10 volts and the current increased (stepped up) to 2 amperes. Notice in the above case that the power remains the same. That is, 20 volts times 1 ampere equals the same power as 10 volts times 2 amperes.

Q1. What is meant by "transformer action?"

BASIC OPERATION OF A TRANSFORMER

In its most basic form a transformer consists of:

- A primary coil or winding.

- A secondary coil or winding.

- A core that supports the coils or windings.

Refer to the transformer circuit in figure 5-1 as you read the following explanation: The primary winding is connected to a 60 hertz ac voltage source. The magnetic field (flux) builds up (expands) and collapses (contracts) about the primary winding. The expanding and contracting magnetic field around the primary winding cuts the secondary winding and induces an alternating voltage into the winding. This voltage causes alternating current to flow through the load. The voltage may be stepped up or down depending on the design of the primary and secondary windings.

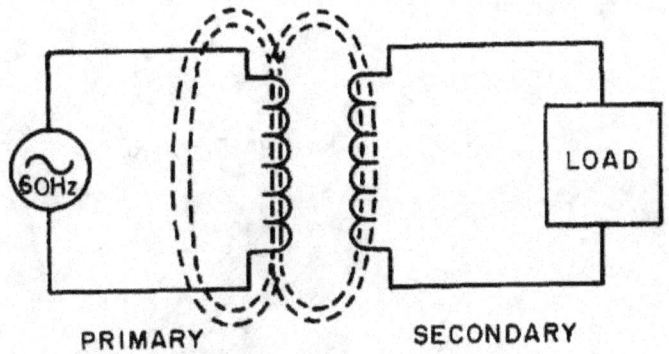

Figure 5-1.—Basic transformer action.

Q2. What are, the three basic parts of a transformer?

THE COMPONENTS OF A TRANSFORMER

Two coils of wire (called windings) are wound on some type of core material. In some cases the coils of wire are wound on a cylindrical or rectangular cardboard form. In effect, the core material is air and the transformer is called an AIR-CORE TRANSFORMER. Transformers used at low frequencies, such as 60 hertz and 400 hertz, require a core of low-reluctance magnetic material, usually iron. This type of transformer is called an IRON-CORE TRANSFORMER. Most power transformers are of the iron-core type. The principle parts of a transformer and their functions are:

- The CORE, which provides a path for the magnetic lines of flux.

- The PRIMARY WINDING, which receives energy from the ac source.

- The SECONDARY WINDING, which receives energy from the primary winding and delivers it to the load.

- The ENCLOSURE, which protects the above components from dirt, moisture, and mechanical damage.

CORE CHARACTERISTICS

The composition of a transformer core depends on such factors as voltage, current, and frequency. Size limitations and construction costs are also factors to be considered. Commonly used core materials are air, soft iron, and steel. Each of these materials is suitable for particular applications and unsuitable for others. Generally, air-core transformers are used when the voltage source has a high frequency (above 20 kHz). Iron-core transformers are usually used when the source frequency is low (below 20 kHz). A soft-iron-core transformer is very useful where the transformer must be physically small, yet efficient. The iron-core transformer provides better power transfer than does the air-core transformer. A transformer whose core is constructed of laminated sheets of steel dissipates heat readily; thus it provides for the efficient transfer of power. The majority of transformers you will encounter in Navy equipment contain laminated-steel cores. These steel laminations (see figure 5-2) are insulated with a nonconducting material, such as varnish, and then formed into a core. It takes about 50 such laminations to make a core an inch thick. The purpose of the laminations is to reduce certain losses which will be discussed later in this chapter. An important point to remember is that the most efficient transformer core is one that offers the best path for the most lines of flux with the least loss in magnetic and electrical energy.

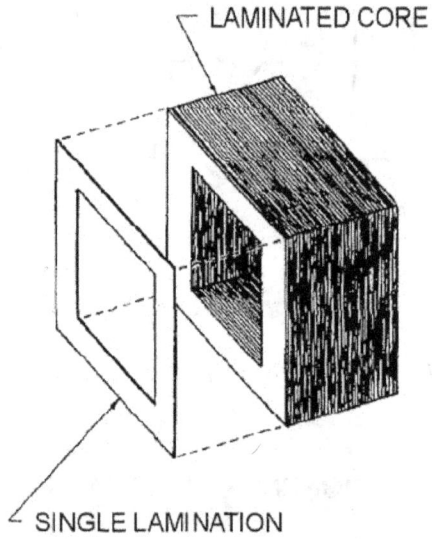

Figure 5-2.—Hollow-core construction.

Q3. What are three materials commonly used as the core of a transformer?

Hollow-Core Transformers

There are two main shapes of cores used in laminated-steel-core transformers. One is the HOLLOW-CORE, so named because the core is shaped with a hollow square through the center. Figure 5-2 illustrates this shape of core. Notice that the core is made up of many laminations of steel. Figure 5-3 illustrates how the transformer windings are wrapped around both sides of the core.

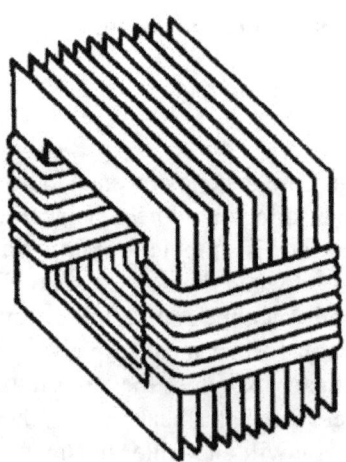

Figure 5-3.—Windings wrapped around laminations.

Shell-Core Transformers

The most popular and efficient transformer core is the SHELL CORE, as illustrated in figure 5-4. As shown, each layer of the core consists of E- and I-shaped sections of metal. These sections are butted together to form the laminations. The laminations are insulated from each other and then pressed together to form the core.

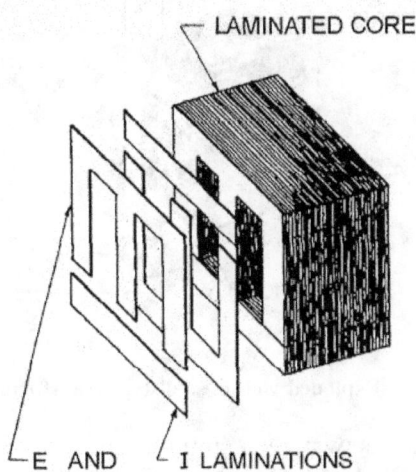

Figure 5-4.—Shell-type core construction.

Q4. What are the two main types of cores used in transformers?

TRANSFORMER WINDINGS

As stated above, the transformer consists of two coils called WINDINGS which are wrapped around a core. The transformer operates when a source of ac voltage is connected to one of the windings and a load device is connected to the other. The winding that is connected to the source is called the PRIMARY WINDING. The winding that is connected to the load is called the SECONDARY WINDING. (Note: In this chapter the terms "primary winding" and "primary" are used interchangeably; the term: "secondary winding" and "secondary" are also used interchangeably.)

Figure 5-5 shows an exploded view of a shell-type transformer. The primary is wound in layers directly on a rectangular cardboard form.

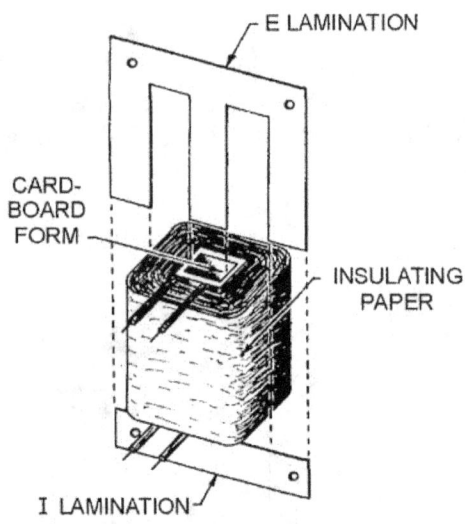

Figure 5-5.—Exploded view of shell-type transformer construction.

In the transformer shown in the cutaway view in figure 5-6, the primary consists of many turns of relatively small wire. The wire is coated with varnish so that each turn of the winding is insulated from every other turn. In a transformer designed for high-voltage applications, sheets of insulating material, such as paper, are placed between the layers of windings to provide additional insulation.

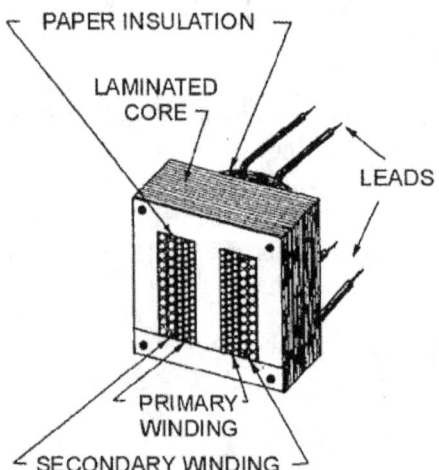

Figure 5-6.—Cutaway view of shell-type core with windings.

When the primary winding is completely wound, it is wrapped in insulating paper or cloth. The secondary winding is then wound on top of the primary winding. After the secondary winding is complete, it too is covered with insulating paper. Next, the E and I sections of the iron core are inserted into and around the windings as shown.

The leads from the windings are normally brought out through a hole in the enclosure of the transformer. Sometimes, terminals may be provided on the enclosure for connections to the windings. The figure shows four leads, two from the primary and two from the secondary. These leads are to be connected to the source and load, respectively.

Q5. Which transformer windings are connected to an ac source voltage and to a load, respectively?

Q6. A transformer designed for high-voltage applications differs in construction in what way from a transformer designed for low-voltage applications?

SCHEMATIC SYMBOLS FOR TRANSFORMERS

Figure 5-7 shows typical schematic symbols for transformers. The symbol for an air-core transformer is shown in figure 5-7(A). Parts (B) and (C) show iron-core transformers. The bars between the coils are used to indicate an iron core. Frequently, additional connections are made to the transformer windings at points other than the ends of the windings. These additional connections are called TAPS. When a tap is connected to the center of the winding, it is called a CENTER TAP. Figure 5-7(C) shows the schematic representation of a center-tapped iron-core transformer.

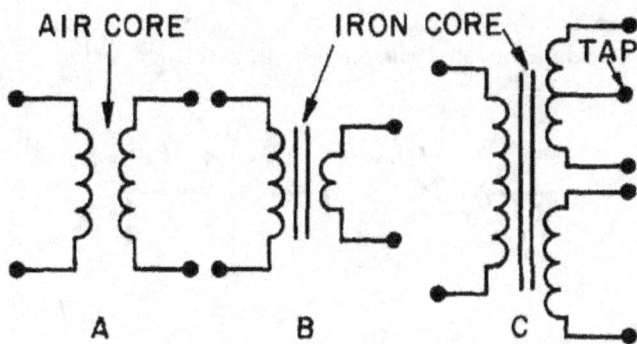

Figure 5-7.—Schematic symbols for various types of transformers.

Q7. Identify the below schematic symbols of transformers by labeling them in the blanks provided.

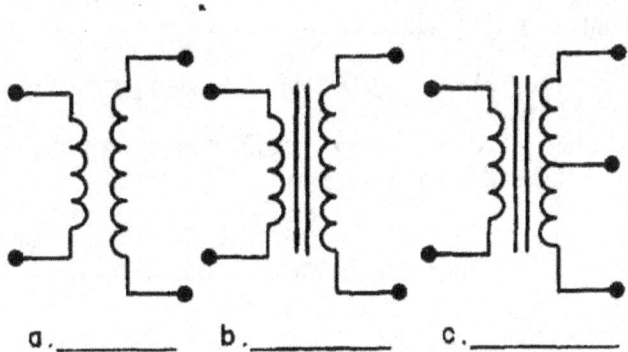

5-7

HOW A TRANSFORMER WORKS

Up to this point the chapter has presented the basics of the transformer including transformer action, the transformer's physical characteristics, and how the transformer is constructed. Now you have the necessary knowledge to proceed into the theory of operation of a transformer.

NO-LOAD CONDITION

You have learned that a transformer is capable of supplying voltages which are usually higher or lower than the source voltage. This is accomplished through mutual induction, which takes place when the changing magnetic field produced by the primary voltage cuts the secondary winding.

A no-load condition is said to exist when a voltage is applied to the primary, but no load is connected to the secondary, as illustrated by figure 5-8. Because of the open switch, there is no current flowing in the secondary winding. With the switch open and an ac voltage applied to the primary, there is, however, a very small amount of current called EXCITING CURRENT flowing in the primary. Essentially, what the exciting current does is "excite" the coil of the primary to create a magnetic field. The amount of exciting current is determined by three factors: (1) the amount of voltage applied (E_a), (2) the resistance (R) of the primary coil's wire and core losses, and (3) the X_L which is dependent on the frequency of the exciting current. These last two factors are controlled by transformer design.

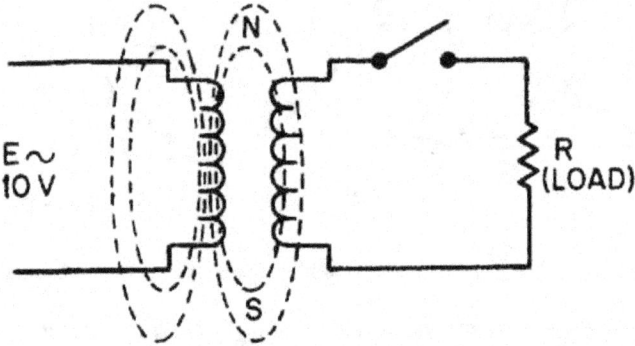

Figure 5-8.—Transformer under no-load conditions.

This very small amount of exciting current serves two functions:

1. Most of the exciting energy is used to maintain the magnetic field of the primary.

2. A small amount of energy is used to overcome the resistance of the wire and core losses which are dissipated in the form of heat (power loss).

Exciting current will flow in the primary winding at all times to maintain this magnetic field, but no transfer of energy will take place as long as the secondary circuit is open.

Q8. What is meant by a "no-load condition" in a transformer circuit?

PRODUCING A COUNTER EMF

When an alternating current flows through a primary winding, a magnetic field is established around the winding. As the lines of flux expand outward, relative motion is present, and a counter emf is induced in the winding. This is the same counter emf that you learned about in the chapter on inductors. Flux leaves the primary at the north pole and enters the primary at the south pole. The counter emf induced in the primary has a polarity that opposes the applied voltage, thus opposing the flow of current in the primary. It is the counter emf that limits exciting current to a very low value.

Q9. What is meant by "exciting current" in a transformer?

INDUCING A VOLTAGE IN THE SECONDARY

To visualize how a voltage is induced into the secondary winding of a transformer, again refer to figure 5-8. As the exciting current flows through the primary, magnetic lines of force are generated. During the time current is increasing in the primary, magnetic lines of force expand outward from the primary and cut the secondary. As you remember, a voltage is induced into a coil when magnetic lines cut across it. Therefore, the voltage across the primary causes a voltage to be induced across the secondary.

Q10. What is the name of the emf generated in the primary that opposes the flow of current in the primary?

Q11. What causes a voltage to be developed across the secondary winding of a transformer?

PRIMARY AND SECONDARY PHASE RELATIONSHIP

The secondary voltage of a simple transformer may be either in phase or out of phase with the primary voltage. This depends on the direction in which the windings are wound and the arrangement of the connections to the external circuit (load). Simply, this means that the two voltages may rise and fall together or one may rise while the other is falling.

Transformers in which the secondary voltage is in phase with the primary are referred to as LIKE-WOUND transformers, while those in which the voltages are 180 degrees out of phase are called UNLIKE-WOUND transformers.

Dots are used to indicate points on a transformer schematic symbol that have the same instantaneous polarity (points that are in phase).

The use of phase-indicating dots is illustrated in figure 5-9. In part (A) of the figure, both the primary and secondary windings are wound from top to bottom in a clockwise direction, as viewed from above the windings. When constructed in this manner, the top lead of the primary and the top lead of the secondary have the SAME polarity. This is indicated by the dots on the transformer symbol. A lack of phasing dots indicates a reversal of polarity.

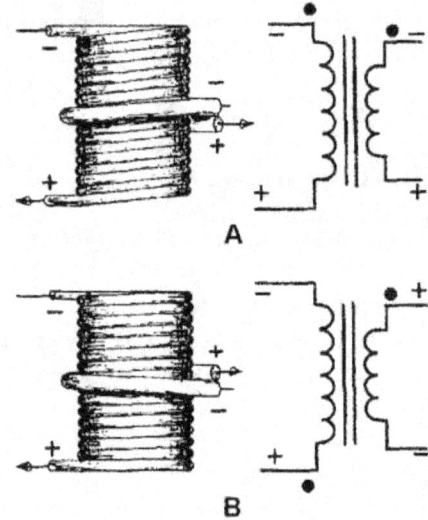

Figure 5-9.—Instantaneous polarity depends on direction of winding.

Part (B) of the figure illustrates a transformer in which the primary and secondary are wound in opposite directions. As viewed from above the windings, the primary is wound in a clockwise direction from top to bottom, while the secondary is wound in a counterclockwise direction. Notice that the top leads of the primary and secondary have OPPOSITE polarities. This is indicated by the dots being placed on opposite ends of the transformer symbol. Thus, the polarity of the voltage at the terminals of the secondary of a transformer depends on the direction in which the secondary is wound with respect to the primary.

Q12. What is the phase relationship between the voltage induced in the secondary of an unlike-wound transformer and the counter emf of the primary winding?

Q13. Draw dots on the below symbol to indicate the phasing of the transformer.

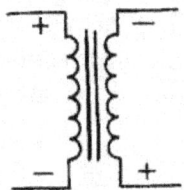

COEFFICIENT OF COUPLING

The COEFFICIENT OF COUPLING of a transformer is dependent on the portion of the total flux lines that cuts both primary and secondary windings. Ideally, all the flux lines generated by the primary should cut the secondary, and all the lines of the flux generated by the secondary should cut the primary. The coefficient of coupling would then be one (unity), and maximum energy would be transferred from the primary to the secondary. Practical power transformers use high-permeability silicon steel cores and close spacing between the windings to provide a high coefficient of coupling.

Lines of flux generated by one winding which do not link with the other winding are called LEAKAGE FLUX. Since leakage flux generated by the primary does not cut the secondary, it cannot induce a voltage into the secondary. The voltage induced into the secondary is therefore less than it would be if the leakage flux did not exist. Since the effect of leakage flux is to lower the voltage induced into the secondary, the effect can be duplicated by assuming an inductor to be connected in series with the primary. This series LEAKAGE INDUCTANCE is assumed to drop part of the applied voltage, leaving less voltage across the primary.

Q14. What is "leakage flux?"

Q15. What effect does flux leakage in a transformer have on the coefficient of coupling (K) in the transformer?

TURNS AND VOLTAGE RATIOS

The total voltage induced into the secondary winding of a transformer is determined mainly by the RATIO of the number of turns in the primary to the number of turns in the secondary, and by the amount of voltage applied to the primary. Refer to figure 5-10. Part (A) of the figure shows a transformer whose primary consists of ten turns of wire and whose secondary consists of a single turn of wire. You know that as lines of flux generated by the primary expand and collapse, they cut BOTH the ten turns of the primary and the single turn of the secondary. Since the length of the wire in the secondary is approximately the same as the length of the wire in each turn in the primary, EMF INDUCED INTO THE SECONDARY WILL BE THE SAME AS THE EMF INDUCED INTO EACH TURN IN THE PRIMARY. This means that if the voltage applied to the primary winding is 10 volts, the counter emf in the primary is almost 10 volts. Thus, each turn in the primary will have an induced counter emf of approximately one-tenth of the total applied voltage, or one volt. Since the same flux lines cut the turns in both the secondary and the primary, each turn will have an emf of one volt induced into it. The transformer in part (A) of figure 5-10 has only one turn in the secondary, thus, the emf across the secondary is one volt.

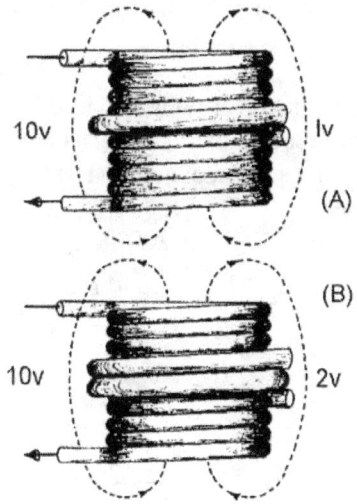

Figure 5-10.—Transformer turns and voltage ratios.

The transformer represented in part (B) of figure 5-10 has a ten-turn primary and a two-turn secondary. Since the flux induces one volt per turn, the total voltage across the secondary is two volts. Notice that the volts per turn are the same for both primary and secondary windings. Since the counter emf in the primary is equal (or almost) to the applied voltage, a proportion may be set up to express the value of the voltage induced in terms of the voltage applied to the primary and the number of turns in each winding. This proportion also shows the relationship between the number of turns in each winding and the voltage across each winding. This proportion is expressed by the equation:

$$\frac{E_S}{E_P} = \frac{N_S}{N_P}$$

Where:
N_P = number of turns in the primary
E_P = voltage applied to the primary
E_S = voltage induced in the secondary
N_S = number of turns in the secondary

Notice the equation shows that the ratio of secondary voltage to primary voltage is equal to the ratio of secondary turns to primary turns. The equation can be written as:

$$E_P N_S = E_S N_P$$

The following formulas are derived from the above equation:

Transposing for E_S: $\quad E_S = \dfrac{E_P N_S}{N_P}$

Transposing for E_P: $\quad E_P = \dfrac{E_S N_P}{N_S}$

If any three of the quantities in the above formulas are known, the fourth quantity can be calculated. Example. A transformer has 200 turns in the primary, 50 turns in the secondary, and 120 volts applied to the primary (E_p). What is the voltage across the secondary (E_S)?

Given:
N_P = 200 turns
N_S = 50 turns
E_P = 120 volts
E_S = ? volts

Solution: $\quad E_S = \dfrac{E_P N_S}{N_P}$

Substitution: $\quad E_S = \dfrac{120 \text{ volts} \times 50 \text{ turns}}{200 \text{ turns}}$

E_S = 30 volts

Example. There are 400 turns of wire in an iron-core coil. If this coil is to be used as the primary of a transformer, how many turns must be wound on the coil to form the secondary winding of the transformer to have a secondary voltage of one volt if the primary voltage is five volts?

Given: $N_P = 400$ turns
$E_P = 5$ volts
$E_S = 1$ volt
$N_S = ?$ turns

Solution: $E_P N_S = E_S N_P$

Transposing for N_S:

$$N_S = \frac{E_S N_P}{E_P}$$

Substitution: $N_S = \dfrac{1 \text{ volt} \times 400 \text{ turns}}{5 \text{ volts}}$

$N_S = 80$ turns

Note: The ratio of the voltage (5:1) is equal to the turns ratio (400:80). Sometimes, instead of specific values, you are given a turns or voltage ratio. In this case, you may assume any value for one of the voltages (or turns) and compute the other value from the ratio. For example, if a turn ratio is given as 6:1, you can assume a number of turns for the primary and compute the secondary number of turns (60:10, 36:6, 30:5, etc.).

The transformer in each of the above problems has fewer turns in the secondary than in the primary. As a result, there is less voltage across the secondary than across the primary. A transformer in which the voltage across the secondary is less than the voltage across the primary is called a STEP-DOWN transformer. The ratio of a four-to-one step-down transformer is written as 4:1. A transformer that has fewer turns in the primary than in the secondary will produce a greater voltage across the secondary than the voltage applied to the primary. A transformer in which the voltage across the secondary is greater than the voltage applied to the primary is called a STEP-UP transformer. The ratio of a one-to-four step-up transformer should be written as 1:4. Notice in the two ratios that the value of the primary winding is always stated first.

Q16. Does 1:5 indicate a step-up or step-down transformer?

Q17. A transformer has 500 turns on the primary and 1500 turns on the secondary. If 45 volts are applied to the primary, what is the voltage developed across the secondary? (Assume no losses)

Q18. A transformer has a turns ratio of 7:1. If 5 volts is developed across the secondary, what is the voltage applied to the primary? (Note: E_S is given, what is E_P?)

Q19. A transformer has 60 volts applied to its primary and 420 volts appearing across its secondary. If there are 800 turns on the primary, what is the number of turns in the secondary?

EFFECT OF A LOAD

When a load device is connected across the secondary winding of a transformer, current flows through the secondary and the load. The magnetic field produced by the current in the secondary interacts with the magnetic field produced by the current in the primary. This interaction results from the mutual inductance between the primary and secondary windings.

MUTUAL FLUX

The total flux in the core of the transformer is common to both the primary and secondary windings. It is also the means by which energy is transferred from the primary winding to the secondary winding. Since this flux links both windings, it is called MUTUAL FLUX. The inductance which produces this flux is also common to both windings and is called mutual inductance.

Figure 5-11 shows the flux produced by the currents in the primary and secondary windings of a transformer when source current is flowing in the primary winding.

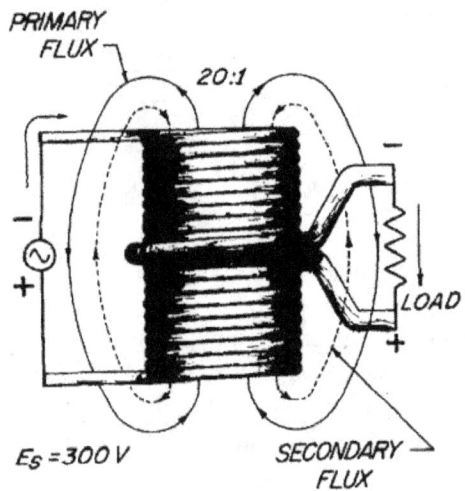

Figure 5-11.—Simple transformer indicating primary- and secondary-winding flux relationship.

When a load resistance is connected to the secondary winding, the voltage induced into the secondary winding causes current to flow in the secondary winding. This current produces a flux field about the secondary (shown as broken lines) which is in opposition to the flux field about the primary (Lenz's law). Thus, the flux about the secondary cancels some of the flux about the primary. With less flux surrounding the primary, the counter emf is reduced and more current is drawn from the source. The additional current in the primary generates more lines of flux, nearly reestablishing the original number of total flux lines.

TURNS AND CURRENT RATIOS

The number of flux lines developed in a core is proportional to the magnetizing force (IN AMPERE-TURNS) of the primary and secondary windings. The ampere-turn ($I \times N$) is a measure of magnetomotive force; it is defined as the magnetomotive force developed by one ampere of current flowing in a coil of one turn. The flux which exists in the core of a transformer surrounds both the primary and secondary windings. Since the flux is the same for both windings, the ampere-turns in both the primary and secondary windings must be the same.

Therefore:

$$I_P N_P = I_S N_S$$

Where:
$I_P N_P$ = ampere – turns in the primary winding
$I_S N_S$ = ampere – turns in the secondary winding

By dividing both sides of the equation by $I_P N_S$, you obtain:

$$\frac{N_P}{N_S} = \frac{I_S}{I_P}$$

Since:
$$\frac{E_S}{E_P} = \frac{N_S}{N_P}$$

Then:
$$\frac{E_P}{E_S} = \frac{N_P}{N_S}$$

And:
$$\frac{E_P}{E_S} = \frac{I_S}{I_P}$$

Where:
E_P = voltage applied to the primary in volts
E_S = voltage across the secondary in volts
I_P = current in the primary in amperes
I_S = current in the secondary in amperes

Notice the equations show the current ratio to be the inverse of the turns ratio and the voltage ratio. This means, a transformer having less turns in the secondary than in the primary would step down the voltage, but would step up the current. Example: A transformer has a 6:1 voltage ratio. Find the current in the secondary if the current in the primary is 200 milliamperes.

Given: $E_P = 6\,V$ (assumed)
$E_S = 1\,V$
$I_P = 200\,mA$ or $0.2\,A$
$I_S = ?$

Solution: $\dfrac{E_P}{E_S} = \dfrac{I_S}{I_P}$

Transposing for I_S:

$$I_S = \dfrac{E_P I_P}{E_S}$$

Substitution:

$$I_S = \dfrac{6\,V \times 0.2\,A}{1\,V}$$

$$I_S = 1.2\,A$$

The above example points out that although the voltage across the secondary is one-sixth the voltage across the primary, the current in the secondary is six times the current in the primary.

The above equations can be looked at from another point of view. The expression

$$\dfrac{N_p}{N_s}$$

is called the transformer TURNS RATIO and may be expressed as a single factor. Remember, the turns ratio indicates the amount by which the transformer increases or decreases the voltage applied to the primary. For example, if the secondary of a transformer has two times as many turns as the primary, the voltage induced into the secondary will be two times the voltage across the primary. If the secondary has one-half as many turns as the primary, the voltage across the secondary will be one-half the voltage across the primary. However, the turns ratio and the current ratio of a transformer have an inverse relationship. Thus, a 1:2 step-up transformer will have one-half the current in the secondary as in the primary. A 2:1 step-down transformer will have twice the current in the secondary as in the primary.

Example: A transformer with a turns ratio of 1:12 has 3 amperes of current in the secondary. What is the value of current in the primary?

Given: $N_P = 1\,turn$ (assumed)
$N_S = 12\,turns$
$I_S = 3\,A$
$I_P = ?$

Solution: $\dfrac{N_P}{N_S} = \dfrac{I_S}{I_P}$

Transposing for I_P:

$$I_P = \frac{N_S I_S}{N_P}$$

Substitution:

$$I_P = \frac{12 \text{ turns} \times 3\text{A}}{1 \text{ turn}}$$

$$I_P = 36 \text{ A}$$

Q20. *A transformer with a turns ratio of 1:3 has what current ratio?*

Q21. *A transformer has a turns ratio of 5:1 and a current of 5 amperes flowing in the secondary. What is the current flowing in the primary? (Assume no losses)*

POWER RELATIONSHIP BETWEEN PRIMARY AND SECONDARY WINDINGS

As just explained, the turns ratio of a transformer affects current as well as voltage. If voltage is doubled in the secondary, current is halved in the secondary. Conversely, if voltage is halved in the secondary, current is doubled in the secondary. In this manner, all the power delivered to the primary by the source is also delivered to the load by the secondary (minus whatever power is consumed by the transformer in the form of losses). Refer again to the transformer illustrated in figure 5-11. The turns ratio is 20:1. If the input to the primary is 0.1 ampere at 300 volts, the power in the primary is $P = E \times I = 30$ watts. If the transformer has no losses, 30 watts is delivered to the secondary. The secondary steps down the voltage to 15 volts and steps up the current to 2 amperes. Thus, the power delivered to the load by the secondary is $P = E \times I = 15$ volts $\times$ 2 amps = 30 watts.

The reason for this is that when the number of turns in the secondary is decreased, the opposition to the flow of the current is also decreased. Hence, more current will flow in the secondary. If the turns ratio of the transformer is increased to 1:2, the number of turns on the secondary is twice the number of turns on the primary. This means the opposition to current is doubled. Thus, voltage is doubled, but current is halved due to the increased opposition to current in the secondary. The important thing to remember is that with the exception of the power consumed within the transformer, all power delivered to the primary by the source will be delivered to the load. The form of the power may change, but the power in the secondary almost equals the power in the primary.

As a formula:

$$P_S = P_P - P_L$$

Where:

P_S = power delivered to the load by the secondary

P_P = power delivered to the primary by the source

P_L = power losses in the transformer

TRANSFORMER LOSSES

Practical power transformers, although highly efficient, are not perfect devices. Small power transformers used in electrical equipment have an 80 to 90 percent efficiency range, while large, commercial powerline transformers may have efficiencies exceeding 98 percent.

The total power loss in a transformer is a combination of three types of losses. One loss is due to the dc resistance in the primary and secondary windings. This loss is called COPPER loss or I^2R loss. The two other losses are due to EDDY CURRENTS and to HYSTERESIS in the core of the transformer. Copper loss, eddy-current loss, and hysteresis loss result in undesirable conversion of electrical energy into heat energy.

Q22. What is the mathematical relationship between the power in the primary (P_P) and power in the secondary (P_S) of a transformer?

Copper Loss

Whenever current flows in a conductor, power is dissipated in the resistance of the conductor in the form of heat. The amount of power dissipated by the conductor is directly proportional to the resistance of the wire, and to the square of the current through it. The greater the value of either resistance or current, the greater is the power dissipated. The primary and secondary windings of a transformer are usually made of low-resistance copper wire. The resistance of a given winding is a function of the diameter of the wire and its length. Copper loss can be minimized by using the proper diameter wire. Large diameter wire is required for high-current windings, whereas small diameter wire can be used for low-current windings.

Eddy-Current Loss

The core of a transformer is usually constructed of some type of ferromagnetic material because it is a good conductor of magnetic lines of flux.

Whenever the primary of an iron-core transformer is energized by an alternating-current source, a fluctuating magnetic field is produced. This magnetic field cuts the conducting core material and induces a voltage into it. The induced voltage causes random currents to flow through the core which dissipates power in the form of heat. These undesirable currents are called EDDY CURRENTS.

To minimize the loss resulting from eddy currents, transformer cores are LAMINATED. Since the thin, insulated laminations do not provide an easy path for current, eddy-current losses are greatly reduced.

Hysteresis Loss

When a magnetic field is passed through a core, the core material becomes magnetized. To become magnetized, the domains within the core must align themselves with the external field. If the direction of the field is reversed, the domains must turn so that their poles are aligned with the new direction of the external field.

Power transformers normally operate from either 60 Hz, or 400 Hz alternating current. Each tiny domain must realign itself twice during each cycle, or a total of 120 times a second when 60 Hz alternating current is used. The energy used to turn each domain is dissipated as heat within the iron core. This loss, called HYSTERESIS LOSS, can be thought of as resulting from molecular friction. Hysteresis loss can be held to a small value by proper choice of core materials.

TRANSFORMER EFFICIENCY

To compute the efficiency of a transformer, the input power to and the output power from the transformer must be known. The input power is equal to the product of the voltage applied to the primary and the current in the primary. The output power is equal to the product of the voltage across the secondary and the current in the secondary. The difference between the input power and the output power represents a power loss. You can calculate the percentage of efficiency of a transformer by using the standard efficiency formula shown below:

$$\text{Efficiency (in \%)} = \frac{P_{out}}{P_{in}} \times 100$$

Where:
P_{out} = total output power delivered to the load
P_{in} = total input power

Example. If the input power to a transformer is 650 watts and the output power is 610 watts, what is the efficiency?

Solution:

$$\text{Efficiency} = \frac{P_{out}}{P_{in}} \times 100$$

$$\text{Efficiency} = \frac{610\,W}{650\,W} \times 100$$

$$\text{Efficiency} = 93.8\%$$

Hence, the efficiency is approximately 93.8 percent, with approximately 40 watts being wasted due to heat losses.

Q23. *Name the three power losses in a transformer.*

Q24. *The input power to a transformer is 1,000 watts and the output power is 500 watts. What is the efficiency of the transformer, expressed as a percentage?*

TRANSFORMER RATINGS

When a transformer is to be used in a circuit, more than just the turns ratio must be considered. The voltage, current, and power-handling capabilities of the primary and secondary windings must also be considered.

The maximum voltage that can safely be applied to any winding is determined by the type and thickness of the insulation used. When a better (and thicker) insulation is used between the windings, a higher maximum voltage can be applied to the windings.

The maximum current that can be carried by a transformer winding is determined by the diameter of the wire used for the winding. If current is excessive in a winding, a higher than ordinary amount of power will be dissipated by the winding in the form of heat. This heat may be sufficiently high to cause the insulation around the wire to break down. If this happens, the transformer may be permanently damaged.

The power-handling capacity of a transformer is dependent upon its ability to dissipate heat. If the heat can safely be removed, the power-handling capacity of the transformer can be increased. This is sometimes accomplished by immersing the transformer in oil, or by the use of cooling fins. The power-handling capacity of a transformer is measured in either the volt-ampere unit or the watt unit.

Two common power generator frequencies (60 hertz and 400 hertz) have been mentioned, but the effect of varying frequency has not been discussed. If the frequency applied to a transformer is increased, the inductive reactance of the windings is increased, causing a greater ac voltage drop across the windings and a lesser voltage drop across the load. However, an increase in the frequency applied to a transformer should not damage it. But, if the frequency applied to the transformer is decreased, the reactance of the windings is decreased and the current through the transformer winding is increased. If the decrease in frequency is enough, the resulting increase in current will damage the transformer. For this reason a transformer may be used at frequencies above its normal operating frequency, but not below that frequency.

Q25. Why should a transformer designed for 400 hertz operation not be used for 60 hertz operation?

TYPES AND APPLICATIONS OF TRANSFORMERS

The transformer has many useful applications in an electrical circuit. A brief discussion of some of these applications will help you recognize the importance of the transformer in electricity and electronics.

POWER TRANSFORMERS

Power transformers are used to supply voltages to the various circuits in electrical equipment. These transformers have two or more windings wound on a laminated iron core. The number of windings and the turns per winding depend upon the voltages that the transformer is to supply. Their coefficient of coupling is 0.95 or more.

You can usually distinguish between the high-voltage and low-voltage windings in a power transformer by measuring the resistance. The low-voltage winding usually carries the higher current and therefore has the larger diameter wire. This means that its resistance is less than the resistance of the high-voltage winding, which normally carries less current and therefore may be constructed of smaller diameter wire.

So far you have learned about transformers that have but one secondary winding. The typical power transformer has several secondary windings, each providing a different voltage. The schematic symbol for a typical power-supply transformer is shown in figure 5-12. For any given voltage across the primary, the voltage across each of the secondary windings is determined by the number of turns in each secondary. A winding may be center-tapped like the secondary 350 volt winding shown in the figure. To center tap a winding means to connect a wire to the center of the coil, so that between this center tap and either terminal of the winding there appears one-half of the voltage developed across the entire winding. Most power transformers have colored leads so that it is easy to distinguish between the various windings to which they are connected. Carefully examine the figure which also illustrates the color code for a typical power transformer. Usually, red is used to indicate the high-voltage leads, but it is possible for a manufacturer to use some other color(s).

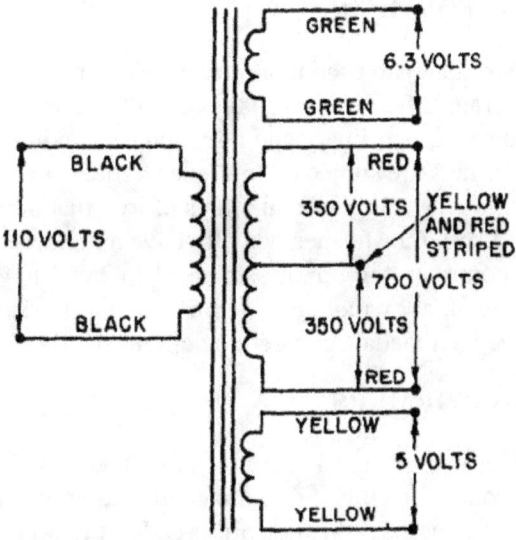

Figure 5-12.—Schematic diagram of a typical power transformer.

There are many types of power transformers. They range in size from the huge transformers weighing several tons-used in power substations of commercial power companies-to very small ones weighing as little as a few ounces-used in electronic equipment.

AUTOTRANSFORMERS

It is not necessary in a transformer for the primary and secondary to be separate and distinct windings. Figure 5-13 is a schematic diagram of what is known as an AUTOTRANSFORMER. Note that a single coil of wire is "tapped" to produce what is electrically a primary and secondary winding. The voltage across the secondary winding has the same relationship to the voltage across the primary that it would have if they were two distinct windings. The movable tap in the secondary is used to select a value of output voltage, either higher or lower than E_P, within the range of the transformer. That is, when the tap is at point A, E_S is less than E_P; when the tap is at point B, E_S is greater than E_P.

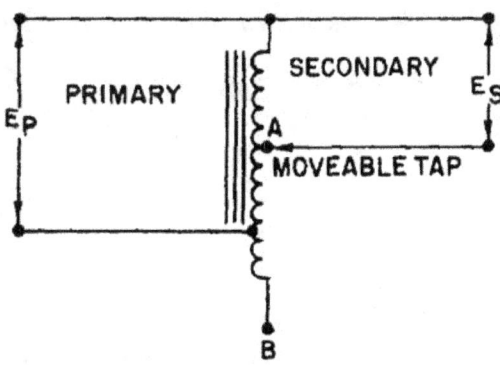

Figure 5-13.—Schematic diagram of an autotransformer.

AUDIO-FREQUENCY TRANSFORMERS

Audio-frequency (af) transformers are used in af circuits as coupling devices. Audio-frequency transformers are designed to operate at frequencies in the audio frequency spectrum (generally considered to be 15 Hz to 20kHz). They consist of a primary and a secondary winding wound on a laminated iron or steel core. Because these transformers are subjected to higher frequencies than are power transformers, special grades of steel such as silicon steel or special alloys of iron that have a very low hysteresis loss must be used for core material. These transformers usually have a greater number of turns in the secondary than in the primary; common step-up ratios being 1 to 2 or 1 to 4. With audio transformers the impedance of the primary and secondary windings is as important as the ratio of turns, since the transformer selected should have its impedance match the circuits to which it is connected.

RADIO-FREQUENCY TRANSFORMERS

Radio-frequency (rf) transformers are used to couple circuits to which frequencies above 20,000 Hz are applied. The windings are wound on a tube of nonmagnetic material, have a special powdered-iron core, or contain only air as the core material. In standard broadcast radio receivers, they operate in a frequency range of from 530 kHz to 1550 kHz. In a short-wave receiver, rf transformers are subjected to frequencies up to about 20 MHz - in radar, up to and even above 200 MHz.

IMPEDANCE-MATCHING TRANSFORMERS

For maximum or optimum transfer of power between two circuits, it is necessary for the impedance of one circuit to be matched to that of the other circuit. One common impedance-matching device is the transformer. To obtain proper matching, you must use a transformer having the correct turns ratio. The number of turns on the primary and secondary windings and the impedance of the transformer have the following mathematical relationship:

$$\frac{N_P}{N_S} = \sqrt{\frac{Z_P}{Z_S}}$$

Because of this ability to match impedances, the impedance-matching transformer is widely used in electronic equipment.

Q26. *List five different types of transformers according to their applications.*

Q27. *The leads to the primary and to the high-voltage secondary windings of a power transformer usually are of what color?*

SAFETY

EFFECTS OF CURRENT ON THE BODY

Before learning safety precautions, you should look at some of the possible effects of electrical current on the human body. The following table lists some of the probable effects of electrical current on the human body.

AC 60 Hz (mA)	DC (mA)	Effects
0-1	0-4	Perception
1-4	4-15	Surprise
4-21	15-80	Reflex action
21-40	80-160	Muscular inhibition
40-100	160-300	Respiratory failure
Over 100	Over 300	Usually fatal

Note in the above chart that a current as low as 4 mA can be expected to cause a reflex action in the victim, usually causing the victim to jump away from the wire or other component supplying the current. While the current should produce nothing more than a tingle of the skin, the quick action of trying to get away from the source of this irritation could produce other effects (such as broken limbs or even death if a severe enough blow was received at a vital spot by the shock victim).

It is important for you to recognize that the resistance of the human body cannot be relied upon to prevent a fatal shock from a voltage as low as 115 volts or even less. Fatalities caused by human contact with 30 volts have been recorded. Tests have shown that body resistance under unfavorable conditions may be as low as 300 ohms, and possibly as low as 100 ohms (from temple to temple) if the skin is broken. Generally direct current is not considered as dangerous as an equal value of alternating current. This is evidenced by the fact that reasonably safe "let-go currents" for 60 hertz, alternating current, are 9.0 milliamperes for men and 6.0 milliamperes for women, while the corresponding values for direct current are 62.0 milliamperes for men and 41.0 milliamperes for women. Remember, the above table is a fist of probable effects. The actual severity of effects will depend on such things as the physical condition of the work area, the physiological condition and resistance of the body, and the area of the body through which the current flows. Thus, based on the above information, you MUST consider every voltage as being dangerous.

ELECTRIC SHOCK

Electric shock is a jarring, shaking sensation you receive from contact with electricity. You usually feel like you have received a sudden blow. If the voltage and resulting current are sufficiently high, you may become unconscious. Severe burns may appear on your skin at the place of contact; muscular spasms may occur, perhaps causing you to clasp the apparatus or wire which caused the shock and be unable to turn it loose.

RESCUE AND CARE OF SHOCK VICTIMS

The following procedures are recommended for rescue and care of electric shock victims:

1. Remove the victim from electrical contact at once, but DO NOT endanger yourself. You can do this by:

   - Throwing the switch if it is nearby

   - Cutting the cable or wires to the apparatus, using an ax with a wooden handle while taking care to protect your eyes from the flash when the wires are severed

- Using a dry stick, rope, belt, coat, blanket, shirt or any other nonconductor of electricity, to drag or push the victim to safety

2. Determine whether the victim is breathing. If the victim is not breathing, you must apply artificial ventilation (respiration) without delay, even though the victim may appear to be lifeless. DO NOT STOP ARTIFICIAL RESPIRATION UNTIL MEDICAL AUTHORITY PRONOUNCES THE VICTIM DEAD.

3. Lay the victim face up. The feet should be about 12 inches higher than the head. Chest or head injuries require the head to be slightly elevated. If there is vomiting or if facial injuries have occurred which cause bleeding into the throat, the victim should be placed on the stomach with the head turned to one side and 6 to 12 inches lower than the feet.

4. Keep the victim warm. The injured person's body heat must be conserved. Keep the victim covered with one or more blankets, depending on the weather and the person's exposure to the elements. Artificial means of warming, such as hot water bottles should not be used.

5. Drugs, food, and liquids should not be administered if medical attention will be available within a short time. If necessary, liquids may be administered. Small amounts of warm salt water, tea or coffee should be used. Alcohol, opiates, and other depressant substances must never be administered.

6. Send for medical personnel (a doctor if available) at once, but do NOT under any circumstances leave the victim until medical help arrives.

For complete coverage of administering artificial respiration, and on treatment of burn and shock victims, refer to *Standard First Aid Training Course*, NAVEDTRA 10081 (Series).

SAFETY PRECAUTIONS FOR PREVENTING ELECTRIC SHOCK

You must observe the following safety precautions when working on electrical equipment:

1. Never work alone. Another person may save your life if you receive an electric shock.

2. Work on energized circuits ONLY WHEN ABSOLUTELY NECESSARY. Power should be tagged out, using approved tagout procedures, at the nearest source of electricity.

3. Stand on an approved insulating material, such as a rubber mat.

4. Discharge power capacitors before working on deenergized equipment. Remember, a capacitor is an electrical power storage device.

5. When you must work on an energized circuit, wear rubber gloves and cover as much of your body as practical with an insulating material (such as shirt sleeves). This is especially important when you are working in a warm space where sweating may occur.

6. Deenergize equipment prior to hooking up or removing test equipment.

7. Work with only one hand inside the equipment. Keep the other hand clear of all obstacles that may provide a path, such as a ground, for current to flow.

8. Wear safety goggles. Sparks could damage your eyes, as could the cooling liquids in some components such as transformers should they overheat and explode.

9. Keep a cool head and think about the possible consequences before performing any action. Carelessness is the cause of most accidents. Remember the best technician is NOT necessarily the fastest one, but the one who will be on the job tomorrow.

Q28. What is the cause of most accidents?

Q29. Before working on electrical equipment containing capacitors, what should you do to the capacitors?

Q30. When working on electrical equipment, why should you use only one hand?

SUMMARY

As a study aid and for future reference, the important points of this chapter have been summarized below.

BASIC TRANSFORMER—The basic transformer is an electrical device that transfers alternating-current energy from one circuit to another circuit by magnetic coupling of the primary and secondary windings of the transformer. This is accomplished through mutual inductance (M). The coefficient of coupling (K) of a transformer is dependent upon the size and shape of the coils, their relative positions, and the characteristic of the core between the two coils. An ideal transformer is one where all the magnetic lines of flux produced by the primary cut the entire secondary. The higher the K of the transformer, the higher is the transfer of the energy. The voltage applied to the primary winding causes current to flow in the primary. This current generates a magnetic field, generating a counter emf (cemf) which has the opposite phase to that of the applied voltage. The magnetic field generated by the current in the primary also cuts the secondary winding and induces a voltage in this winding.

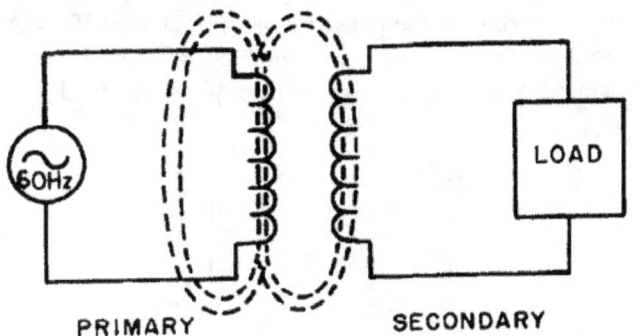

TRANSFORMER CONSTRUCTION—A TRANSFORMER consists of two coils of insulated wire wound on a core. The primary winding is usually wound onto a form, then wrapped with an insulating material such as paper or cloth. The secondary winding is then wound on top of the primary and both windings are wrapped with insulating material. The windings are then fitted onto the core of the transformer. Cores come in various shapes and materials. The most common materials are air, soft iron, and laminated steel. The most common types of transformers are the shell-core and the hollow-core types. The type and shape of the core is dependent on the intended use of the transformer and the voltage applied to the current in the primary winding.

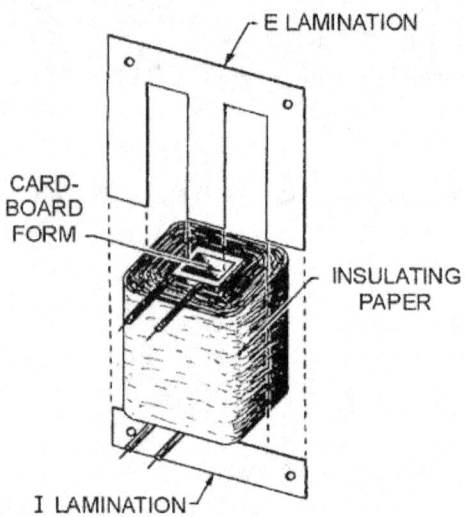

EXCITING CURRENT—When voltage is applied to the primary of a transformer, exciting current flows in the primary. The current causes a magnetic field to be set up around both the primary and the secondary windings. The moving flux causes a voltage to be induced into the secondary winding, countering the effects of the counter emf in the primary.

PHASE—When the secondary winding is connected to a load, causing current to flow in the secondary, the magnetic field decreases momentarily. The primary then draws more current, restoring the magnetic field to almost its original magnitude. The phase of the current flowing in the secondary circuit is dependent upon the phase of the voltage impressed across the primary and the direction of the winding of the secondary. If the secondary were wound in the same direction as the primary, the phase would be the same. If wound opposite to the primary, the phase would be reversed. This is shown on a schematic drawing by the use of phasing dots. The dots mean that the leads of the primary and secondary have the same phase. The lack of phasing dots on a schematic means a phase reversal.

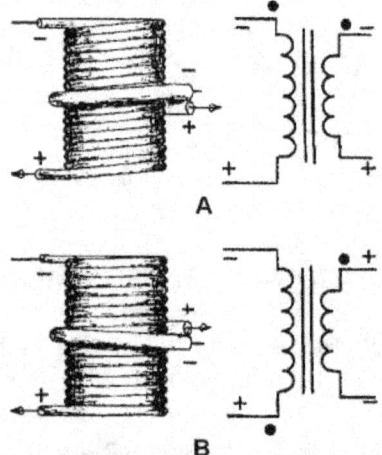

TURNS RATIO—The TURNS RATIO of a transformer is the ratio of the number of turns of wire in the primary winding to the number of turns in the secondary winding. When the turns ratio is stated, the number representing turns on the primary is always stated first. For example, a 1:2 turns ratio means the secondary has twice the number of turns as the primary. In this example, the voltage across the secondary is two times the voltage applied to the primary.

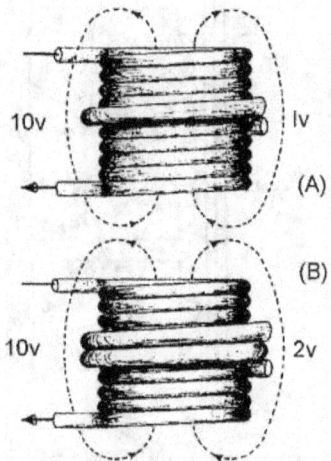

POWER AND CURRENT RATIOS—The power and current ratios of a transformer are dependent on the fact that power delivered to the secondary is always equal to the power delivered to the primary minus the losses of the transformer. This will always be true, regardless of the number of secondary windings. Using the law of power and current, it can be stated that current through the transformer is the inverse of the voltage or turns ratio, with power remaining the same or less, regardless of the number of secondaries.

TRANSFORMER LOSSES—Transformer losses have two sources-copper loss and magnetic loss. Copper losses are caused by the resistance of the wire (I^2R). Magnetic losses are caused by eddy currents and hysteresis in the core. Copper loss is a constant after the coil has been wound and therefore a measureable loss. Hysteresis loss is constant for a particular voltage and current. Eddy-current loss, however, is different for each frequency passed through the transformer.

TRANSFORMER EFFICIENCY—The amplitude of the voltage induced in the secondary is dependent upon the efficiency of the transformer and the turns ratio. The efficiency of a transformer is related to the power losses in the windings and core of the transformer. Efficiency (in percent) equals $P_{out}/P_{in} \times 100$. A perfect transformer would have an efficiency of 1.0 or 100%.

POWER TRANSFORMER—A transformer with two or more windings wound on a laminated iron core. The transformer is used to supply stepped up and stepped down values of voltage to the various circuits in electrical equipment.

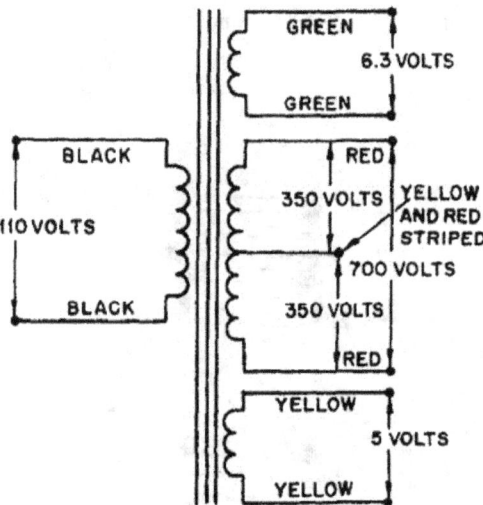

AUTOTRANSFORMER—A transformer with a single winding in which the entire winding can be used as the primary and part of the winding as the secondary, or part of the winding can be used as the primary and the entire winding can be used as the secondary.

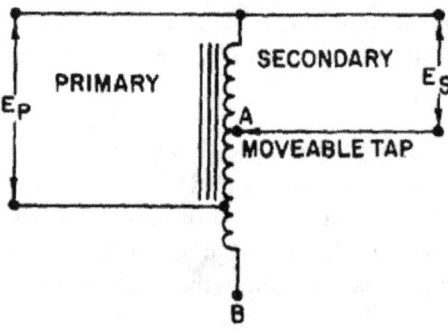

AUDIO-FREQUENCY TRANSFORMER—A transformer used in audio-frequency circuits to transfer af signals from one circuit to another.

RADIO-FREQUENCY TRANSFORMER—A transformer used in a radio-frequency circuit to transfer rf signals from one circuit to another.

IMPEDANCE-MATCHING TRANSFORMER—A transformer used to match the impedance of the source and the impedance of the load. The mathematical relationship of the turns and impedance of the transformer is expressed by the equation:

$$\frac{N_P}{N_S} = \sqrt{\frac{Z_P}{Z_S}}$$

ANSWERS TO QUESTIONS Q1. THROUGH Q30.

A1. *The transfer of energy from one circuit to another circuit by electromagnetic induction.*

A2. *Primary winding; secondary winding; core.*

A3. *Air; soft iron; steel.*

A4. *Hollow-core type; shell-core type.*

A5. *Primary to source; secondary to load.*

A6. *Additional insulation is provided between the layers of windings in the high-voltage transformer.*

A7.

    a. air-core transformer

    b. iron-core transformer

    c. iron-core center tapped transformer

A8. *A voltage is applied to the primary, but no load is connected to the secondary.*

A9. *Exciting current is the current that flows in the primary of a transformer with the secondary open (no load attached).*

A10. *Self-induced or counter emf.*

A11. *The magnetic lines generated by the current in the primary cut the secondary windings and induce a voltage into them.*

A12. *In phase. Remember, the cemf of the primary is 180 degrees out of phase with the applied voltage. The induced voltage of the secondary of an unlike-wound transformer is also 180 degrees out of phase with the primary voltage.*

A13.

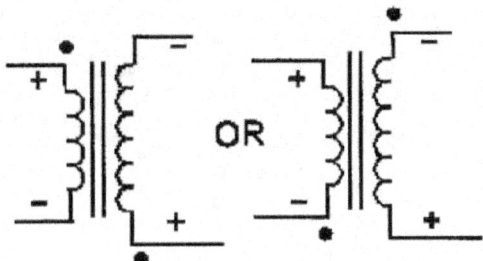

Note: Remember the dots indicate areas of like polarity, NOT a particular polarity.

A14. *Lines of flux generated by one winding which do not link the other winding.*

A15. *It causes K to be less than unity (1).*

A16. *Step up.*

5-29

A17.

$$\frac{E_S}{E_P} = \frac{N_S}{N_P} \text{ or}$$

$$E_S = \frac{E_P N_S}{N_P} = \frac{45\text{ V} \times 1500\text{ turns}}{500\text{ turns}} = 135\text{ V}$$

A18.

$$\frac{E_P}{E_S} = \frac{N_P}{N_S} \text{ or}$$

$$E_P = \frac{N_P E_S}{N_S} = \frac{7\text{ turns} \times 5\text{ V}}{1\text{ turn}} = 35\text{ V}$$

A19.

$$\frac{E_P}{E_S} = \frac{N_P}{N_S} \text{ or}$$

$$N_S = \frac{E_S N_P}{E_P} = \frac{420\text{ V} \times 800\text{ turns}}{60\text{ V}} = 5600\text{ turns}$$

A20.

$$\frac{N_P}{N_S} = \frac{I_S}{I_P} = \frac{1}{3} = 3:1 \text{ current ratio}$$

(Turns ratio and current ratio have an inverse relationship.)

A21.

$$\frac{N_P}{N_S} = \frac{I_S}{I_P} \text{ or}$$

$$I_P = \frac{N_S I_S}{N_P} = \frac{1\text{ turn} \times 5\text{ A}}{5\text{ turns}} = 1\text{ A}$$

A22.

$$P_S = P_P - P_L$$

A23. *Copper loss, eddy-current loss, and hysteresis loss.*

A24.

$$\text{Eff (in \%)} = \frac{P_{out}}{P_{in}} \times 100 = \frac{500\,W}{1000\,W} \times 100 = 0.5 \times 100 = 50\%$$

A25. The inductive reactance at 60 hertz would be too low. The resulting excessive current would probably damage the transformer.

A26.
    a. Power transformer
    b. Autotransformer
    c. Impedance-matching transformer
    d. Audio-frequency transformer
    e. Radio-frequency transformer

A27. Primary leads-black; secondary leads-red.

A28. Carelessness.

A29. Discharge them by shorting them to ground.

A30. To minimize the possibility of providing a path for current through your body.

APPENDIX I
GLOSSARY

AIR-CORE TRANSFORMER—A transformer composed of two or more coils, which are wound around a non-metallic core.

ALTERNATING CURRENT—An electrical current which constantly changes amplitude and changes polarity at regular intervals.

APPARENT POWER—That power apparently available for use in an ac circuit containing a reactive element. It is the product of effective voltage times effective current expressed in voltamperes. It must be multiplied by the power factor to obtain true power available.

AVERAGE VALUE OF AC—The average of all the instantaneous values of one-half cycle of alternating current.

CAPACITANCE—The property of an electrical circuit that opposes changes in voltage.

CAPACITOR—An electrical device capable of storing electrical energy in an electrostatic field.

CAPACITIVE REACTANCE—The opposition offered to the flow of an alternating current by capacitance, expressed in ohms. The symbol for capacitive reactance is X_c.

CHARGE CYCLE—The period of time that a capacitor in an electrical circuit is storing a charge.

COIL—An inductive device created by looping turns of wire around a core.

COPPER LOSS (I^2R LOSS)—The power lost due to the resistance of the conductors. In transformers the power lost due to heating because of current flow (I) through the resistance (R) of the windings.

CORE—Any material that affords a path for magnetic flux lines in a coil.

COUNTER EMF—Counter electromotive force; an electromotive force (voltage) induced in a coil that opposes the applied voltage.

COUPLING, COEFFICIENT OF—An expression of the extent to which two inductors are coupled by magnetic lines of force. This is expressed as a decimal or percentage of maximum possible coupling and represented by the letter K.

CYCLE—One complete positive and one complete negative alternation of a current or voltage.

DIELECTRIC—An insulator; a term applied to the insulating material between the plates of a capacitor.

DIELECTRIC CONSTANT—The ratio of the capacitance of a capacitor with a dielectric between the electrodes to the capacitance of a capacitor with air between the electrodes.

DIELECTRIC HYSTERESIS LOSS—Power loss of a capacitor due to the changes in orientation of electron orbits in the dielectric caused by rapid reversal in polarity of line voltage. The higher the frequency, the greater the loss.

DIELECTRIC LEAKAGE—Power loss of a capacitor due to the leakage of current through the dielectric. Also relates to leakage resistance, the higher the leakage resistance, the lower the dielectric leakage.

DISPLACEMENT CURRENT—The current which appears to flow through a capacitor.

EDDY CURRENT—Induced circulating currents in a conducting material that are caused by a varying magnetic field.

EDDY CURRENT LOSS—Losses caused by random current flowing in the core of a transformer. Power is lost in the form of heat.

EFFECTIVE VALUE—Same as root-mean-square.

ELECTROMAGNETIC INDUCTION—The production of a voltage in a coil due to the change in the number of magnetic lines of force (flux linkage) passing through the coil.

ELECTROMAGNETISM—The generation of a magnetic field around a current carrying conductor.

ELECTROMOTIVE FORCE (emf)—The force (voltage) that produces an electric current in a circuit

ELECTROSTATIC FIELD—The field of influence between two charged bodies.

EXCITING CURRENT—The current that flows in the primary winding of a transformer, which produces a magnetic flux field. Also called magnetizing current.

FARAD—The basic unit of capacitance. A capacitor has a capacitance of 1 farad when a voltage change of 1 volt per second across it produces a current of 1 ampere.

FLUX—Electrostatic or magnetic lines of force.

FREQUENCY (f)—The number of complete cycles per second existing in any form of wave motion; such as the number of cycles per second of an alternating current.

HENRY (H)—The electromagnetic unit of inductance or mutual inductance. The inductance of a circuit is 1 henry when a current variation of 1 ampere per second induces 1 volt. It is the basic unit of inductance. In radio, smaller units are used such as the millihenry (mH), which is one-thousandth of a henry (H) and the microhenry (μH) hich is onemillionth of a henry.

HERTZ (Hz)—The basic unit of frequency equal to one cycle per second.

HYSTERESIS—The time lag of the magnetic flux in a magnetic material behind the magnetizing force producing it, caused by the molecular friction of the molecules trying to align themselves with the magnetic force applied to the material.

HYSTERESIS LOSS—The power loss in an iron-core transformer or other alternating-current device as a result of magnetic hysteresis.

IMPEDANCE—The total opposition offered to the flow of an alternating current. It may consist of any combination of resistance, inductive reactance, and capacitive reactance. The symbol for impedance is Z.

IN PHASE—Applied to the condition that exists when two waves of the same frequency pass through their maximum and minimum values of like polarity at the same instant.

INDUCTANCE—The property of a circuit which tends to oppose a change in the existing current flow. The symbol for inductance is L.

INDUCTIVE COUPLING—Coupling of two coils by means of magnetic lines of force. In transformers, coupling applied through magnetic lines of force between the primary and secondary windings.

INDUCTIVE REACTANCE—The opposition to the flow of an alternating current caused by the inductance of a circuit, expressed in ohms. Identified by the letter X_L.

INSTANTANEOUS VALUE—The magnitude at any particular instant when a value is continually varying with respect to time.

LAG—The amount one wave is behind another in time; expressed in electrical degrees.

LAMINATED CORE—A core built up from thin sheets of metal insulated from each other and used in transformers.

LEAD—The opposite of lag. Also a wire or connection.

LEAKAGE FLUX—Magnetic flux lines produced by the primary winding which do not link the turns of the secondary winding.

LEAKAGE RESISTANCE—The electrical resistance which opposes the flow of current through the dielectric of a capacitor. The higher the leakage resistance the slower the capacitor will discharge or leak across the dielectric.

LEFT-HAND-RULE FOR GENERATORS—A rule or procedure used to determine the direction of electron current flow in the windings of a generator.

LENZ'S LAW—The current induced in a circuit due to its motion in a magnetic field or to a change in its magnetic flux in such a direction as to exert a mechanical force opposing the motion or to oppose the change in flux.

MAGNETIC FIELD—Region in which the magnetic forces created by a permanent magnet or by a current-carrying conductor or coil can be detected.

MAGNETIC LINES OF FORCE—Imaginary lines used for convenience to designate the direction in which magnetic forces are acting as a result of magneto-motive force.

MUTUAL FLUX—The total flux in the core of a transformer that is common to both the primary and secondary windings. The flux links both windings.

MUTUAL INDUCTANCE—A circuit property existing when the relative position of two inductors causes the magnetic lines of force from one to link with the turns of the other. The symbol for mutual inductance is M.

NEGATIVE ALTERNATION—The negative half of an ac waveform

NO-LOAD CONDITION—The condition that exists when an electrical source or secondary of a transformer is operated without an electrical load.

PEAK-TO-PEAK—The measure of absolute magnitude of an ac waveform, measured from the greatest positive alternation to greatest negative alternation.

PEAK VALUE—The maximum instantaneous value of a varying current, voltage, or power. It is equal to 1.414 times the effective value of a sine wave.

PERIOD TIME—The time required to complete one cycle of a waveform.

PHASE—The angular relationship between two alternating currents or voltages when the voltage or current is plotted as a function of time. When the two are in phase, the angle is zero, and both reach their peak simultaneously. When out of phase, one will lead or lag the other; at the instant when one is at its peak, the other will not be at peak value and (depending on the phase angle) may differ in polarity as well as magnitude.

PHASE ANGLE—The number of electrical degrees of lead or lag between the voltage and current waveforms in an ac circuit.

PHASE DIFFERENCE—The time in electrical degrees by which one wave leads or lags another.

POLARIZATION—The magnetic orientation of molecules in a magnetizable material in a magnetic field, whereby tiny internal magnets tend to line up in the field.

POSITIVE ALTERNATION—The positive half of an ac waveform.

POWER FACTOR—The ratio of the actual power of an alternating or pulsating current, as measured by a wattmeter, to the apparent power, as indicated by ammeter and voltmeter readings. The power factor of an inductor, capacitor, or insulator is an expression of their losses

POWER LOSS—The electrical power supplied to a circuit that does no work, usually dissipated as heat.

PRIMARY WINDING—The winding of a transformer connected to the electrical source.

Q—Figure of merit of efficiency of a circuit or coil. Ratio of inductive reactance to resistance in servos. Relationship between stored energy (capacitance) and rate of dissipation in certain types of electric elements, structures, or materials.

RADIO FREQUENCY (RF)—Any frequency of electrical energy capable of propagation into space.

RATIO—The value obtained by dividing one number by another, indicating their relative proportions.

RC CONSTANT—Time constant of a resistor-capacitor circuit; equal in seconds to the resistance value in ohms multiplied by the capacitance value in farads.

REACTANCE—The opposition offered to the flow of an alternating current by the inductance, capacitance, or both, in any circuit.

RESONANCE—The condition existing in a circuit when the values of inductance, capacitance, and the applied frequency are such that the inductive reactance and capacitive reactance cancel each other.

RLC CIRCUIT—An electrical circuit which has the properties of resistance, inductance, and capacitance.

RMS—Abbreviation of Root Mean Square.

ROOT MEAN SQUARE (RMS)—The equivalent heating value of an alternating current or voltage, as compared to a direct current or voltage. It is 0.707 times the peak value of a sine wave.

SECONDARY—The output coil of a transformer.

SELF-INDUCTION—The production of a counter-electromotive force in a conductor when its own magnetic field collapses or expands with a change in current in the conductor

SINE WAVE—The curve traced by the projection on a uniform time scale of the end of a rotating arm, or vector. Also known as a sinusoidal wave.

THETA—The greek letter (θ) used to represent phase angle.

TIME CONSTANT—The time required to charge a capacitor to 63.2 percent of maximum voltage or discharge to 36.8 percent of its final voltage. The time required for the current in an inductor to increase to 63.2 percent of maximum current or decay to 36.8 percent of its final current.

TRANSFORMER—A device composed of two or more coils, linked by magnetic lines of force, used to transfer electrical energy from one circuit to another.

TRANSFORMER EFFICIENCY—The ratio of output power to input power, generally expressed as a percentage.

$$\text{Efficiency} = \frac{P_{out}}{P_{in}} \times 100$$

TRANSFORMER, STEP-DOWN—A transformer so constructed that the number of turns in the secondary winding is less than the number of turns in the primary winding. This construction will provide less voltage in the secondary circuit than in the primary circuit.

TRANSFORMER, STEP-UP—A transformer so constructed that the number of turns in the secondary winding is more than the number of turns in the primary winding. This construction will provide more voltage in the secondary circuit than in the primary circuit.

TRUE POWER—The power dissipated in the resistance of the circuit, or the power actually used in the circuit.

TURN—One complete loop of a conductor about a core.

TURNS RATIO—The ratio of number of turns in the primary winding to the number of turns in the secondary winding of a transformer.

UNIVERSAL TIME CONSTANT CHART—A chart used to find the time constant of a circuit if the impressed voltage and the values of R and C or R and L are known.

WAVEFORM—The shape of the wave obtained when instantaneous values of an ac quantity are plotted against time in rectangular coordinates.

WAVELENGTH (λ)—The distance, usually expressed in meters, traveled by a wave during the time interval of one complete cycle. It is equal to the velocity of light divided by the frequency.

WORKING VOLTAGE—The maximum voltage that a capacitor may operate at without the risk of damage.

VAR—Abbreviation for Volt-Amperes Reactive.

VECTOR—A line used to represent both direction and magnitude.

APPENDIX II
GREEK ALPHABET

Name	Upper Case	Lower Case	Designates
Alpha.......	A	α	Angles.
Beta.........	B	β	Angles, flux density.
Gamma.....	Γ	γ	Conductivity.
Delta........	Δ	δ	Variation of a quantity, increment.
Epsilon......	E	ε	Base of natural logarithms (2.71828).
Zeta.........	Z	ζ	Impedance, coefficients, coordinates.
Eta...........	H	η	Hysteresis coefficient, efficiency, magnetizing force.
Theta........	Θ	θ	Phase angle.
Iota...........	I	ι	
Kappa......	K	κ	Dielectric constant, coupling coefficient, susceptibility.
Lambda.....	Λ	λ	Wavelength.
Mu...........	M	μ	Permeability, micro, amplification factor.
Nu............	N	ν	Reluctivity.
Xi.............	Ξ	ξ	
Omicron....	O	o	
Pi.............	Π	π	3.1416
Rho..........	P	ρ	Resistivity.
Sigma.......	Σ	σ	
Tau..........	T	τ	Time constant, time-phase displacement.
Upsilon.....	Y	υ	
Phi...........	Φ	ϕ	Angles, magnetic flux.
Chi...........	X	χ	
Psi...........	Ψ	ψ	Dielectric flux, phase difference.
Omega...	Ω	ω	Ohms (upper case), angular velocity ($2\pi f$) (lower case).

APPENDIX III

SQUARE AND SQUARE ROOTS

N	N²	√N	N	N²	√N	N	N²	√N
1	1	1.000	41	1681	6.4031	81	6561	9.0000
2	4	1.414	42	1764	6.4807	82	6724	9.0554
3	9	1.732	43	1849	6.5574	83	6889	9.1104
4	16	2.000	44	1936	6.6332	84	7056	9.1652
5	25	2.236	45	2025	6.7082	85	7225	9.2195
6	36	2.449	46	2116	6.7823	86	7396	9.2736
7	49	2.646	47	2209	6.8557	87	7569	9.3274
8	64	2.828	48	2304	6.9282	88	7744	9.3808
9	81	3.000	49	2401	7.0000	89	7921	9.4340
10	100	3.162	50	2500	7.0711	90	8100	9.4868
11	121	3.3166	51	2601	7.1414	91	8281	9.5394
12	144	3.4641	52	2704	7.2111	92	8464	9.5917
13	169	3.6056	53	2809	7.2801	93	8649	9.6437
14	196	3.7417	54	2916	7.3485	94	8836	9.6954
15	225	3.8730	55	3025	7.4162	95	9025	9.7468
16	256	4.0000	56	3136	7.4833	96	9216	9.7980
17	289	4.1231	57	3249	7.5498	97	9409	9.8489
18	324	4.2426	58	3364	7.6158	98	9604	9.8995
19	361	4.3589	59	3481	7.6811	99	9801	9.9499
20	400	4.4721	60	3600	7.7460	100	10000	10.0000
21	441	4.5826	61	3721	7.8102	101	10201	10.0499
22	484	4.6904	62	3844	7.8740	102	10404	10.0995
23	529	4.7958	63	3969	7.9373	103	10609	10.1489
24	576	4.8990	64	4096	8.0000	104	10816	10.1980
25	625	5.0000	65	4225	8.0623	105	11025	10.2470
26	676	5.0990	66	4356	8.1240	106	11236	10.2956
27	729	5.1962	67	4489	8.1854	107	11449	10.3441
28	784	5.2915	68	4624	8.2462	108	11664	10.3923
29	841	5.3852	69	4761	8.3066	109	11881	10.4403
30	900	5.4772	70	4900	8.3666	110	12100	10.4881
31	961	5.5678	71	5041	8.4261	111	12321	10.5357
32	1024	5.6569	72	5184	8.4853	112	12544	10.5830
33	1089	5.7447	73	5329	8.5440	113	12769	10.6301
34	1156	5.8310	74	5476	8.6023	114	12996	10.6771
35	1225	5.9161	75	5625	8.6603	115	13225	10.7238
36	1296	6.0000	76	5776	8.7178	116	13456	10.7703
37	1369	6.0828	77	5929	8.7750	117	13689	10.8167
38	1444	6.1644	78	6084	8.8318	118	13924	10.8628
39	1521	6.2450	79	6241	8.8882	119	14161	10.9087
40	1600	6.3246	80	6400	8.9443	120	14400	10.9545

For numbers up to 120. For larger numbers divide into factors smaller than 120.

Examples: $\sqrt{225}$ and $\sqrt{16200}$

$225 = 5 \times 45$
$\sqrt{225} = \sqrt{5} \times \sqrt{45}$
$\sqrt{225} = 2.236 \times 6.7082$
$\sqrt{225} = 15$

$16200 = 100 \times 81 \times 2$
$\sqrt{16200} = \sqrt{100} \times \sqrt{81} \times \sqrt{2}$
$\sqrt{16200} = 10 \times 9 \times 1.414$
$\sqrt{16200} = 127.26$

APPENDIX IV
USEFUL AC FORMULAS

PERIOD TIME (t)

$$t = \frac{1}{f}$$

FREQUENCY (f)

$$f = \frac{1}{t}$$

AVERAGE VOLTAGE OR CURRENT

$$E_{avg} = 0.636 \times E_{max}$$
$$I_{avg} = 0.636 \times I_{max}$$

EFFECTIVE VALUE OF VOLTAGE OR CURRENT

$$E_{eff} = 0.707 \times E_{max}$$
$$I_{eff} = 0.707 \times I_{max}$$

MAXIMUM VOLTAGE OR CURRENT

$$I_{max} = 1.414 \times I_{eff}$$
$$E_{max} = 1.414 \times E_{eff}$$

OHM'S LAW OF AC CIRCUIT CONTAINING ONLY RESISTANCE

$$I_{eff} = \frac{E_{eff}}{R}$$
$$I_{avg} = \frac{E_{avg}}{R}$$
$$I_{max} = \frac{E_{max}}{R}$$

L/R TIME CONSTANT (TC)

$$TC \text{ (in seconds)} = \frac{L \text{ (henrys)}}{R \text{ (ohms)}}$$

MUTUAL INDUCTANCE (M)

$$M = K\sqrt{L_1 L_2}$$

TOTAL INDUCTANCE (L_T) Series without magnetic coupling

$$L_T = L_1 + L_2 + L_3 \ldots L_n$$

with magnetic coupling

$$L_T = L_1 + L_2 \pm 2M$$

TOTAL INDUCTANCE (L_T) PARALLEL
(No magnetic coupling)

$$\frac{1}{L_T} = \frac{1}{L_1} + \frac{1}{L_2} + \frac{1}{L_3} \ldots \frac{1}{L_n}$$

CAPACITANCE (C)

$$C \text{ (farads)} = \frac{Q \text{ (coulombs)}}{E \text{ (volts)}}$$

$$C = 0.2249 \frac{KA}{d}$$

RC TIME CONSTANT (t)

$$t \text{ (in seconds)} = R \text{ (ohms)} \times C \text{ (farads)}$$
$$t \text{ (in seconds)} = R \text{ (Mohms)} \times C \text{ (}\mu F\text{)}$$
$$t \text{ (}\mu s\text{)} = R \text{ (ohms)} \times C \text{ (}\mu F\text{)}$$
$$t \text{ (}\mu s\text{)} = R \text{ (Mohms)} \times C \text{ (pF)}$$

TOTAL CAPACITANCE (C_T) SERIES

$$C_T = \frac{1}{\frac{1}{C_1} + \frac{1}{C_2} \cdots \frac{1}{C_n}}$$

$$C_T = \frac{C_1 \times C_2}{C_1 + C_2}$$

TOTAL CAPACITANCE (C_T) PARALLEL

$$C_T = C_1 + C_2 + C_3 \ldots C_n$$

INDUCTIVE REACTANCE (X_L)

$$X_L = 2\pi f L$$

CAPACITIVE REACTANCE (X_C)

$$X_C = \frac{1}{2\pi f C}$$

IMPEDANCE (Z)

$$Z = \sqrt{R^2 + X^2}$$

OHM'S LAW FOR REACTIVE CIRCUITS

$$I = \frac{E}{X_L} \text{ or } I = \frac{E}{X_C}$$

OHM'S LAW FOR CIRCUITS CONTAINING RESISTANCE AND REACTANCE

$$I = \frac{E}{Z}$$

REACTIVE POWER

$$= I_L^2 X_L - I_C^2 X_C$$
$$= I_C^2 X_C - I_L^2 X_L$$

APPARENT POWER

$$= I_Z^2 Z$$
$$= \sqrt{(\text{true power})^2 + (\text{reactive power})^2}$$

POWER FACTOR (PF)

$$PF = \frac{(I_R)^2 R}{(I_Z)^2 Z}$$

VOLTAGE ACROSS THE SECONDARY (E_s)

$$E_S = \frac{E_P N_S}{N_S}$$

VOLTAGE ACROSS THE PRIMARY (E_p)

$$E_P = \frac{E_S N_P}{N_S}$$

CURRENT ACROSS THE SECONDARY (I_s)

$$I_P = \frac{E_S I_S}{E_P}$$

CURRENT ACROSS THE PRIMARY (I_p)

$$I_P = \frac{E_S I_S}{E_P}$$

TRANSFORMER EFFICIENCY

$$= \frac{P_{out}}{P_{in}} \times 100$$

APPENDIX V
TRIGONOMETRIC FUNCTIONS

In a right triangle, there are several relationships which always hold true. These relationships pertain to the length of the sides of a right triangle, and the way the lengths are affected by the angles between them. An understanding of these relationships, called trigonometric functions, is essential for solving problems in a-c circuits such as power factor, impedance, voltage drops, and so forth.

To be a RIGHT triangle, a triangle must have a "square" corner; one in which there is exactly 90° between two of the sides. Trigonometric functions do not apply to any other type of triangle. This type of triangle is shown in figure V-1.

By use of the trigonometric functions, it is possible to determine the UNKNOWN length of one or more sides of a triangle, or the number of degrees in UNKNOWN angles, depending on what is presently known about the triangle. For instance, if the lengths of any two sides are known, the third side and both angles θ (theta) and Φ (phi) may be determined. The triangle may also be solved if the length of any one side and one of the angles (θ or Φ in fig. V-1) are known.

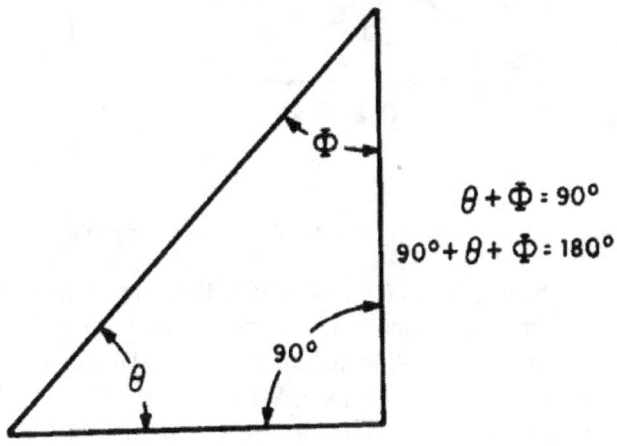

Figure V-1.—A right triangle.

The first basic fact of triangles is that IN ANY RIGHT TRIANGLE, THE SUM OF THE THREE ANGLES FORMED INSIDE THE TRIANGLE MUST ALWAYS EQUAL 180°. If one angle is always 90° (a right angle) then the sum of the other two angles must always be 90°.

$$\theta + \Phi = 90°$$
$$\text{and} \quad 90° + \theta + \Phi = 180°$$
$$\text{or} \quad 90° + 90° = 180°$$

thus, if angle θ is known, Φ may be quickly determined.

For instance, if θ ; is 30° m what is Φ ?

AV-1

$$90° + 30° + \Phi = 180°$$
Transposing $\Phi = 180° - 90° - 30°$
$$\Phi = 60°$$

Also, if θ is known, Φ may be determined in the same manner.

The second basic fact you must understand is that FOR EVERY DIFFERENT COMBINATION OF ANGLES IN A TRIANGLE, THERE IS A DEFINITE RATIO BETWEEN THE LENGTHS OF THE THREE SIDES. Consider the triangle in figure V-2, consisting of the base, side B; the altitude, side A; and the hypotenuse, side C. (The hypotenuse is always the longest side, and is always opposite the 90° angle.)

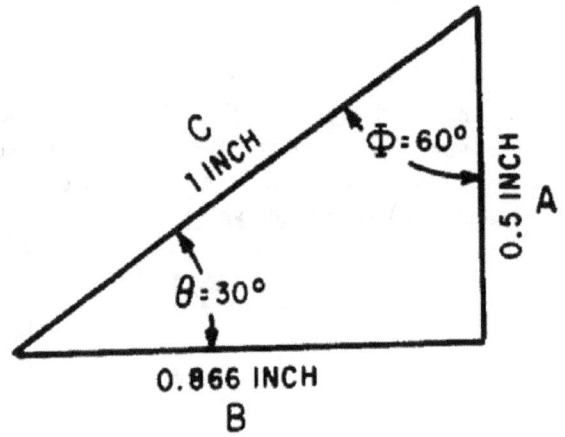

Figure V-2.—A 30° - 60° - 90° triangle.

If angle θ is 30°, Φ must be 60°. With θ equal to 30°, the ratio of the length of side B to side C is 0.866 to 1. That is, if the hypotenuse is 1 inch long, the side adjacent to θ, side B, is 0.866 inch long. Also, with θ equal to 30°, the ratio of side A to side C is 0.5 to 1. That is, with the hypotenuse 1 inch long, the side opposite to θ (side A) is 0.5 inch long. With θ still at 30°, side A is 0.5774 of the length of B. With the combination of angles given (30°-60°-90°) these are the ONLY ratios of lengths that will "fit" to form a right triangle.

Note that three ratios are shown to exist for the given value of θ: the ratio B\C which is always referred to as the COSINE ratio of θ, the ratio A\C, which is always the SINE ratio of θ, and the ratio A\B, which is always the TANGENT ratio of θ. If θ changes, all three ratios change, because the lengths of the sides (base and altitude) change. There is a set of ratios for every increment between 0° and 90°. These angular ratios, or sine, cosine, and tangent functions, are listed for each degree and tenth of degree in a table at the end of this appendix. In this table, the length of the hypotenuse of a triangle is considered fixed. Thus, the ratios of length given refer to the manner in which sides A and B vary with relation to each other and in relation to side C, as angle θ is varied from 0° to 90°.

The solution of problems in trigonometry (solution of triangles is much simpler when the table of trigonometric functions is used properly. The most common ways in which it is used will be shown by solving a series of exemplary problems.

Problem 1: If the hypotenuse of the triangle (side C) in figure V-3 is 10 inches long, and angle θ is 33°, how long are sides B and A?

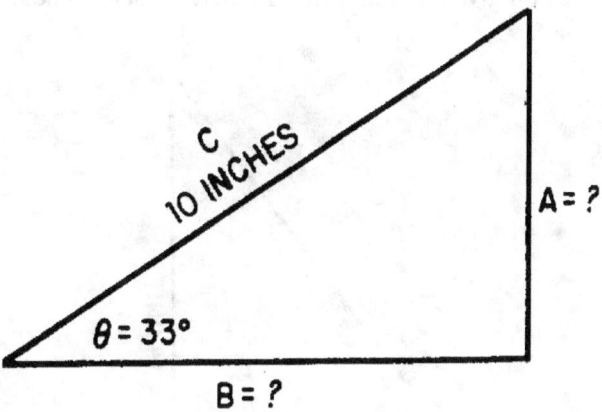

Figure V-3.—Problem 1.

Solution: The ratio B/C is the cosine function. By checking the table of functions, you will find that the cosine of 33° is 0.8387. This means that the length of B is 0.8387 the length of side C. If side C is 10 inches long, then side B must be 10 × 0.8387, or 8.387 inches in length. To determine the length of side A, use the sine function, the ratio A\C. Again consulting the table of functions, you will find that the sine of 33° is 0.5446. Thus, side A must be 10 × 0.5446, or 5.446 inches in length.

Problem 2: The triangle in figure V-4 has a base 74.2 feet long, and hypotenuse 100 feet long. What is θ, and how long is side A?

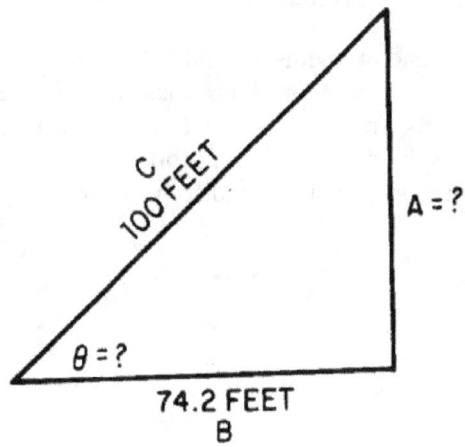

Figure V4.—Problem 2.

Solution: When no angles are given, you Must always solve for a known angle first. The ratio B\C is the cosine of the unknown angle θ; therefore 74.2/100 or 0.742, is the cosine of the unknown angle. Locating 0.742 as a cosine value in the table, you find that it is the cosine of 42.1°. That is, θ = 42.1°. With θ known, side A is solved for by use of the sine ratio A/C. The sine of 42.1°, according to the table, is 0.6704. Therefore, side A is 100 × 0.6704, or 67.04 feet long.

Problem 3: In the triangle in figure V-5, the base is 3 units long, and the altitude is 4 units. What is θ, and how long is the hypotenuse?

Solution: With the information given, the tangent of θ may be determined. Tan θ = A/B = 4/3 = 1.33.

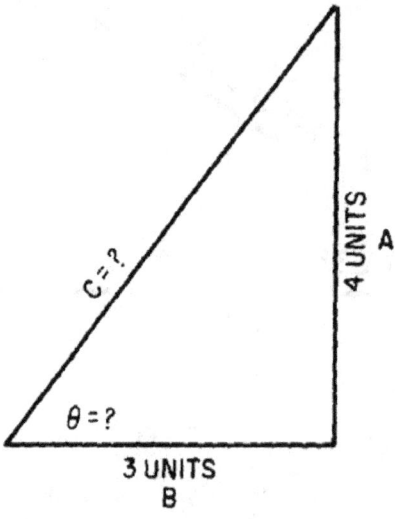

Figure V-5.—Problem 3.

Locating the value 1.33 as a tangent value in the table of functions, you find it to be the tangent of 53.1°. Therefore, θ = 53.1°. Once θ is known, either the sine or cosine ratio may be used to determine the length of the hypotenuse. The cosine of 53.1° is 0.6004. This indicates that the base of 3 units is 0.6004, the length of the hypotenuse. Therefore, the hypotenuse is 3/0.6004, or 5 units in length. Using the sine ratio, the hypotenuse is 4/0.7997, or 5 units in length.

In the foregoing explanations and problems, the sides of triangles were given in inches, feet, and units. In applying trigonometry to a-c circuit problems, these units of measure will be replaced by such values given in ohms, amperes, volts, and watts. Angle θ will be the phase angle between (source) voltage and circuit current. However, the solution of these a-c problems is accomplished in exactly the same manner as the foregoing problems. Only the units and some terminology are changed.

APPENDIX VI
TRIGONOMETRIC TABLES

deg	sin	cos	tan	cot		deg	sin	cos	tan	cot	
0.0	0.0000	1.0000	0.00000		90.0	4.0	0.6976	0.9976	0.6993	140.301	86.0
0.1	0.0175	1.0000	0.00175	573.0	0.9	0.1	0.7150	0.9974	0.7168	130.951	0.9
0.2	0.0349	1.0000	0.00349	286.5	0.8	0.2	0.7324	0.9973	0.7344	130.617	0.8
0.3	0.0524	1.0000	0.00524	191.0	0.7	0.3	0.7498	0.9972	0.17519	130.30	0.7
0.4	0.0698	1.0000	0.00698	143.24	0.6	0.4	0.7672	0.9971	0.7695	120.996	0.6
0.5	0.0873	1.0000	0.00873	114.59	0.5	0.5	0.7846	0.9969	0.7870	120.706	0.5
0.6	0.1047	0.9999	0.10147	95.49	0.4	0.6	0.8020	0.9968	0.8046	120.429	0.4
0.7	0.1222	0.9999	0.01222	81.85	0.3	0.7	0.8194	0.9966	0.8221	120.163	0.3
0.8	0.1396	0.9999	0.01396	71.62	0.2	0.8	0.8368	0.9965	0.8397	110.909	0.2
0.9	0.1571	0.9999	0.01571	63.66	0.1	0.9	0.8542	0.9963	0.8573	110.664	0.1
1.0	0.1745	0.9998	0.1746	570.20	89.0	5.0	0.8716	0.9962	0.8749	110.430	85.0
0.1	0.1920	0.9998	0.1920	520.8	0.9	0.1	0.8889	0.9960	0.89215	110.205	0.9
0.2	0.2094	0.9998	0.2095	470.74	0.8	0.2	0.9063	0.9959	0.9101	10.988	0.8
0.3	0.2269	0.9997	0.2269	440.7	0.7	0.3	0.9237	0.9957	0.9277	10.780	0.7
0.4	0.2443	0.9997	0.2444	40.92	0.6	0.4	0.9411	0.9956	0.9453	10.579	0.6
0.5	0.2618	0.9997	0.2619	380.19	0.5	0.5	0.9585	0.9954	0.9629	10.385	0.5
0.6	0.2792	0.9996	0.2793	350.80	0.4	0.6	0.9758	0.9952	0.9805	10.199	0.4
0.7	0.2967	0.9996	0.2968	330.69	0.4	0.7	0.9932	0.9951	0.9981	10.19	0.3
0.8	0.3141	0.9995	0.3143	310.82	0.2	0.8	0.10106	0.9949	0.10158	90.845	0.2
0.9	0.3316	0.9995	0.3317	30.14	0.1	0.9	0.10279	0.9947	0.10334	90.677	0.1
2.0	0.3490	0.9994	0.3492	280.64	88.0	6.0	0.10453	0.9945	0.10510	90.514	84.0
0.1	0.3664	0.9993	0.3667	270.27	0.9	0.1	0.10626	0.9943	0.10687	90.357	0.9
0.2	0.3839	0.9993	0.3842	260.3	0.8	0.2	0.1080	0.9942	0.10863	90.205	0.8
0.3	0.4013	0.9992	0.4016	240.90	0.7	0.3	0.10973	0.9940	0.11040	90.58	0.7
0.4	0.4188	0.9991	0.4191	230.86	0.6	0.4	0.11147	0.9938	0.11217	80.915	0.6
0.5	0.4362	0.9990	0.4366	220.90	0.5	0.5	0.11320	0.9936	0.11394	80.777	0.5
0.6	0.4536	0.9990	0.4541	220.2	0.4	0.6	0.11494	0.9934	0.11570	80.643	0.4
0.7	0.4711	0.9989	0.4716	210.20	0.3	0.7	0.11667	0.9932	0.11747	80.513	0.3
0.8	0.4885	0.9988	0.4891	20.45	0.2	0.8	0.11840	0.9930	0.11924	80.386	0.2
0.9	0.5059	0.9987	0.5066	190.74	0.1	0.9	0.12014	0.9928	0.12101	80.264	0.1
3.0	0.5234	0.9986	0.5241	190.81	87.0	7.0	0.12187	0.9925	0.12278	80.144	83.0
0.1	0.5408	0.9985	0.5416	180.464	0.9	0.1	0.12360	0.9923	0.12456	80.28	0.9
0.2	0.5582	0.9984	0.5591	170.886	0.8	0.2	0.12533	0.9921	0.12633	70.916	0.8
0.3	0.5756	0.9983	0.5766	170.343	0.7	0.3	0.12706	0.9919	0.12810	70.806	0.7
0.4	0.5931	0.9982	0.5941	160.832	0.6	0.4	0.12880	0.9917	0.12988	70.70	0.6
0.5	0.6105	0.9981	0.6116	160.350	0.5	0.5	0.13053	0.9914	0.13165	70.596	0.5
0.6	0.6279	0.9980	0.6291	150.895	0.4	0.6	0.13226	0.9912	0.13343	70.495	0.4
0.7	0.6453	0.9979	0.6467	150.464	0.3	0.7	0.13399	0.9910	0.13521	70.396	0.3
0.8	0.6627	0.9978	0.6642	150.56	0.2	0.8	0.13572	0.9907	0.13698	70.30	0.2
0.9	0.6802	0.9977	0.6817	140.669	0.1	0.9	0.13744	0.9905	0.13876	70.207	0.1
	cos	sin	cot	tan	deg		cos	sin	cot	tan	

deg	sin	cos	tan	cot		deg	sin	cos	tan	cot	
8.0	0.13917	0.9903	0.14054	70.115	82.0	12.0	0.2079	0.9781	0.2126	40.705	78.0
0.1	0.14090	0.990	0.14232	70.26	0.9	0.1	0.2096	0.9778	0.2144	40.665	0.9
0.2	0.14263	0.9898	0.14410	60.940	0.8	0.2	0.2133	0.9774	0.2162	40.625	0.8
0.3	0.14436	0.9895	0.14588	60.855	0.7	0.3	0.2130	0.9770	0.2180	40.586	0.7
0.4	0.14608	0.9893	0.14767	60.772	0.6	0.4	0.2147	0.9767	0.2199	40.548	0.6
0.5	0.14781	0.9890	0.14945	60.691	0.5	0.5	0.2164	0.9763	0.2217	40.511	0.5
0.6	0.14954	0.9888	0.15124	60.612	0.4	0.6	0.2181	0.9759	0.2235	40.474	0.4
0.7	0.15126	0.9885	0.15302	60.535	0.3	0.7	0.2198	0.9755	0.2254	40.437	0.3
0.8	0.15299	0.9882	0.15481	60.460	0.2	0.8	0.2215	0.9751	0.2272	40.402	0.2
0.9	0.15471	0.9880	0.15660	60.386	0.1	0.9	0.2233	0.9748	0.2290	40.366	0.1
9.0	0.15643	0.9877	0.15836	60.314	81.0	13.0	0.2250	0.9744	0.2309	40.331	77.0
0.1	0.15816	0.9874	0.16017	60.243	0.9	0.1	0.2267	0.9740	0.2327	40.297	0.9
0.2	0.15988	0.9871	0.16196	60.174	0.8	0.2	0.2284	0.9736	0.2345	40.264	0.8
0.3	0.16160	0.9869	0.16376	60.107	0.7	0.3	0.230	0.9732	0.2364	40.230	0.7
0.4	0.16333	0.9866	0.16555	60.41	0.6	0.4	0.2317	0.9728	0.2382	40.198	0.6
0.5	0.16505	0.9863	0.16734	50.976	0.5	0.5	0.2334	0.9724	0.2401	40.165	0.5
0.6	0.16677	0.9860	0.16914	50.912	0.4	0.6	0.2351	0.9720	0.2419	40.134	0.4
0.7	0.16849	0.9857	0.17093	50.850	0.3	0.7	0.2368	0.9715	0.2438	40.102	0.3
0.8	0.17021	0.9854	0.17273	50.789	0.2	0.8	0.2385	0.9711	0.2456	40.71	0.2
0.9	0.17193	0.9851	0.17453	50.730	0.1	0.9	0.2402	0.9707	0.2475	40.41	0.1
10.0	0.1736	0.9848	0.1763	50.671	80.0	14.0	0.2419	0.9703	0.2493	40.11	76.0
0.1	0.1754	0.9845	0.1781	50.614	0.9	0.1	0.2436	0.9699	0.2512	30.981	0.9
0.2	0.1771	0.9842	0.1799	50.558	0.8	0.2	0.2453	0.9694	0.2530	30.952	0.8
0.3	0.1788	0.9839	0.1817	50.503	0.7	0.3	0.2470	0.9680	0.2549	30.923	0.7
0.4	0.1805	0.9836	0.1835	50.449	0.6	0.4	0.2487	0.9686	0.2568	30.895	0.6
0.5	0.1822	0.9833	0.1853	50.396	0.5	0.5	0.2504	0.9681	0.2586	30.867	0.5
0.6	0.1840	0.9829	0.1871	50.343	0.4	0.6	0.2521	0.9677	0.2605	30.839	0.4
0.7	0.1857	0.9826	0.1890	50.292	0.3	0.7	0.2538	0.9673	0.2623	30.812	0.3
0.8	0.1874	0.9823	0.1908	50.242	0.2	0.8	0.2554	0.9668	0.2642	30.785	0.2
0.9	0.1891	0.9820	0.1926	50.193	0.1	0.9	0.2571	0.9664	0.2661	30.758	0.1
11.0	0.1908	0.9816	0.1944	50.145	79.0	15.0	0.2588	0.9659	0.2679	30.732	75.0
0.1	0.1925	0.9813	0.1962	50.97	0.9	0.1	0.2605	0.9655	0.2698	30.706	0.9
0.2	0.1942	0.9810	0.1980	50.50	0.8	0.2	0.2622	0.9650	0.2717	30.681	0.8
0.3	0.1959	0.9806	0.1998	50.5	0.7	0.3	0.2639	0.9646	0.2736	30.655	0.7
0.4	0.1977	0.9803	0.2016	40.959	0.6	0.4	0.2656	0.9641	0.2754	30.630	0.6
0.5	0.1994	0.9799	0.2035	40.915	0.5	0.5	0.2672	0.9636	0.2773	30.606	0.5
0.6	0.2011	0.9796	0.2053	40.872	0.4	0.6	0.2689	0.9632	0.2792	30.582	0.4
0.7	0.2028	0.9792	0.2071	40.829	0.3	0.7	0.2706	0.9627	0.2811	30.558	0.3
0.8	0.2045	0.9789	0.2089	40.787	0.2	0.8	0.2723	0.9622	0.2830	30.534	0.2
0.9	0.2062	0.9785	0.2107	40.745	0.1	0.9	0.2740	0.9617	0.2849	30.511	0.2
	cos	sin	cot	tan	deg		cos	sin	cot	tan	deg

deg	sin	cos	tan	cot		deg	sin	cos	tan	cot	
16.0	0.2756	0.9613	0.2867	30.487	74.0	20.0	0.3420	0.9397	0.3640	20.747	70.0
0.1	0.2773	0.9608	0.2886	30.465	0.9	0.1	0.3437	0.9391	0.3659	20.733	0.9
0.2	0.2790	0.9603	0.2905	30.442	0.8	0.2	0.3453	0.9385	0.3679	20.718	0.8
0.3	0.2807	0.9598	0.2924	30.420	0.7	0.3	0.3469	0.9379	0.3699	20.703	0.7
0.4	0.2823	0.9593	0.2943	30.398	0.6	0.4	0.3486	0.9373	0.3719	20.689	0.6
0.5	0.2840	0.9588	0.2962	30.376	0.5	0.5	0.3502	0.9367	0.3739	20.675	0.5
0.6	0.2857	0.9583	0.2981	30.354	0.4	0.6	0.3518	0.9361	0.3759	20.660	0.4
0.7	0.2874	0.9578	0.300	30.333	0.3	0.7	0.3535	0.9354	0.3779	20.646	0.3
0.8	0.2890	0.9573	0.3019	30.312	0.2	0.8	0.3551	0.9348	0.3799	20.633	0.2
0.9	0.2907	0.9568	0.3038	30.291	0.1	0.9	0.3567	0.9342	0.3819	20.619	0.1
17.0	0.2924	0.9563	0.3067	30.271	73.0	21.0	0.3584	0.9336	0.3839	20.605	69.0
0.1	0.2940	0.9558	0.3076	30.271	0.9	0.1	0.360	0.9330	0.3859	20.592	0.9
0.2	0.2957	0.9553	0.3096	30.230	0.8	0.2	0.3616	0.9323	0.3879	20.578	0.8
0.3	0.2974	0.9548	0.3115	30.211	0.7	0.3	0.3633	0.9317	0.3899	20.565	0.7
0.4	0.2990	0.9542	0.3134	30.191	0.6	0.4	0.3649	0.9311	0.3919	20.552	0.6
0.5	0.307	0.9537	0.3153	30.172	0.5	0.5	0.3665	0.9304	0.3939	20.539	0.5
0.6	0.3024	0.9532	0.3172	30.152	0.4	0.6	0.3681	0.9298	0.3959	20.526	0.4
0.7	0.3040	0.9527	0.3191	30.133	0.3	0.7	0.3697	0.9291	0.3979	20.513	0.3
0.8	0.3057	0.9521	0.3211	30.115	0.2	0.8	0.3714	0.9285	0.400	20.50	0.2
0.9	0.3074	0.9516	0.3230	30.96	0.1	0.9	0.3730	0.9278	0.4020	20.488	0.1
18.0	0.3090	0.9511	0.3249	30.78	72.0	22.0	0.3746	0.9272	0.4040	20.475	68.0
0.1	0.3107	0.9505	0.3269	30.60	0.9	0.1	0.3762	0.9265	0.4061	20.463	0.9
0.2	0.3123	0.950	0.3288	30.42	0.8	0.2	0.3778	0.9259	0.4081	20.450	0.8
0.3	0.3140	0.9494	0.3307	30.24	0.7	0.3	0.3795	0.9252	0.4101	20.438	0.7
0.4	0.3156	0.9489	0.3327	30.6	0.6	0.4	0.3811	0.9245	0.4122	20.426	0.6
0.5	0.3173	0.9483	0.3346	20.989	0.5	0.5	0.3727	0.9239	0.4142	20.414	0.5
0.6	0.3190	0.9478	0.3365	20.971	0.4	0.6	0.3843	0.9232	0.4163	20.402	0.4
0.7	0.3206	0.9472	0.3385	20.954	0.3	0.7	0.3859	0.9225	0.4183	20.391	0.3
0.8	0.3223	0.9466	0.3404	20.937	0.2	0.8	0.3875	0.9219	0.4204	20.379	0.2
0.9	0.3239	0.9461	0.3424	20.921	0.1	0.9	0.3891	0.9212	0.4224	20.367	0.1
19.0	0.3256	0.9455	0.3443	20.904	71.0	23.0	0.3907	0.9205	0.4245	20.356	67.0
0.1	0.3272	0.9449	0.3463	20.888	0.9	0.1	0.3923	0.9198	0.4265	20.344	0.9
0.2	0.3289	0.9444	0.3482	20.872	0.8	0.2	0.3939	0.9191	0.4286	20.333	0.8
0.3	0.3305	0.9438	0.3502	20.856	0.7	0.3	0.3955	0.9184	0.4307	20.322	0.7
0.4	0.3322	0.9432	0.3522	20.840	0.6	0.4	0.3971	0.9178	0.4327	20.311	0.6
0.5	0.3338	0.9426	0.3541	20.824	0.5	0.5	0.3987	0.9171	0.4348	20.30	0.5
0.6	0.3355	0.9421	0.3561	20.808	0.4	0.6	0.403	0.9164	0.4369	20.289	0.4
0.7	0.3371	0.9415	0.3581	20.793	0.3	0.7	0.4019	0.9157	0.4390	20.278	0.3
0.8	0.3387	0.9409	0.360	20.778	0.2	0.8	0.4035	0.9150	0.4411	20.267	0.2
0.9	0.3403	0.9403	0.3620	20.762	0.1	0.9	0.4051	0.9143	0.4431	20.257	0.1
	cos	sin	cot	tan	deg		cos	sin	cot	tan	deg

deg	sin	cos	tan	cot		deg	sin	cos	tan	cot	
24.0	0.4067	0.9135	0.4452	20.246	66.0	28.0	0.4695	0.8829	0.5317	10.881	62.0
0.1	0.4083	0.9128	0.4473	20.236	0.9	0.1	0.4710	0.8821	0.5340	10.873	0.9
0.2	0.4099	0.9121	0.4494	20.225	0.8	0.2	0.4726	0.8813	0.5362	10.865	0.8
0.3	0.4115	0.9114	0.4515	20.215	0.7	0.3	0.4741	0.8805	0.5384	10.857	0.7
0.4	0.4131	0.9107	0.4536	20.204	0.6	0.4	0.4756	0.8796	0.5407	10.849	0.6
0.5	0.4147	0.910	0.4557	20.194	0.5	0.5	0.4772	0.8788	0.5430	10.842	0.5
0.6	0.4163	0.9092	0.4578	20.184	0.4	0.6	0.4787	0.8780	0.5452	10.834	0.4
0.7	0.4179	0.9085	0.4599	20.174	0.3	0.7	0.4802	0.8771	0.5475	10.827	0.3
0.8	0.4195	0.9078	0.4621	20.164	0.2	0.8	0.4818	0.8763	0.5498	10.819	0.2
0.9	0.4210	0.9070	0.4642	20.154	0.1	0.9	0.4833	0.8755	0.5520	10.811	0.1
25.0	0.4226	0.9063	0.4663	20.145	65.0	29.0	0.4848	0.8746	0.5543	10.804	61.0
0.1	0.4242	0.9056	0.4684	20.135	0.9	0.1	0.4863	0.8738	0.5566	10.797	0.9
0.2	0.4258	0.9048	0.4706	20.125	0.8	0.2	0.4879	0.8729	0.5589	10.789	0.8
0.3	0.4274	0.9041	0.4727	20.116	0.7	0.3	0.4894	0.8721	0.5612	10.782	0.7
0.4	0.4289	0.9033	0.4748	20.106	0.6	0.4	0.4909	0.8712	0.5635	10.775	0.6
0.5	0.4305	0.9028	0.4770	20.97	0.5	0.5	0.4924	0.8704	0.5658	10.767	0.5
0.6	0.4321	0.9018	0.4791	20.87	0.4	0.6	0.4939	0.8695	0.5681	10.760	0.4
0.7	0.4337	0.9011	0.4813	20.78	0.3	0.7	0.4955	0.8686	0.5704	10.753	0.3
0.8	0.4352	0.903	0.4834	20.69	0.2	0.8	0.4970	0.8678	0.5726	10.746	0.2
0.9	0.4368	0.8996	0.4856	20.59	0.1	0.9	0.4985	0.8669	0.5750	10.739	0.1
26.0	0.4384	0.8988	0.4877	20.50	64.0	30.0	0.500	0.8660	0.5774	10.7321	60.0
0.1	0.4399	0.8980	0.4899	20.41	0.9	0.1	0.5015	0.8652	0.5797	10.7251	0.9
0.2	0.4415	0.8973	0.4921	20.32	0.8	0.2	0.5030	0.8643	0.5820	10.7162	0.8
0.3	0.4431	0.8965	0.4942	20.23	0.7	0.3	0.5045	0.8634	0.5844	10.7113	0.7
0.4	0.4446	0.8957	0.4964	20.14	0.6	0.4	0.5040	0.8625	0.5867	10.7045	0.6
0.5	0.4462	0.8949	0.4986	20.6	0.5	0.5	0.5075	0.8616	0.5890	10.6977	0.5
0.6	0.4478	0.8942	0.508	10.997	0.4	0.6	0.5090	0.8607	0.5914	10.6909	0.4
0.7	0.4493	0.8934	0.5029	10.988	0.3	0.7	0.5105	0.8599	0.5938	10.6842	0.3
0.8	0.4509	0.8926	0.5051	10.980	0.2	0.8	0.5120	0.8590	0.5961	10.6715	0.2
0.9	0.4524	0.8918	0.5073	10.971	0.1	0.9	0.5135	0.8581	0.5985	10.6709	0.1
27.0	0.4540	0.8910	0.5095	10.963	63.0	31.0	0.5150	0.8572	0.609	10.6643	59.0
0.1	0.4555	0.8902	0.5117	10.954	0.9	0.1	0.5165	0.8643	0.6032	10.6577	0.9
0.2	0.4571	0.8894	0.5139	10.946	0.8	0.2	0.5180	0.8554	0.6056	10.6512	0.8
0.3	0.4586	0.8886	0.5161	10.937	0.7	0.3	0.5195	0.8545	0.6080	10.6447	0.7
0.4	0.4602	0.8878	0.5184	10.929	0.6	0.4	0.5210	0.8536	0.6104	10.6383	0.6
0.5	0.4617	0.8870	0.5206	10.921	0.5	0.5	0.5225	0.8526	0.6128	10.6319	0.5
0.6	0.4633	0.8862	0.5228	10.913	0.4	0.6	0.5240	0.8517	0.6152	10.6255	0.4
0.7	0.4648	0.8854	0.5250	10.905	0.3	0.7	0.5255	0.8508	0.6176	10.6191	0.3
0.8	0.4664	0.8846	0.5272	10.897	0.2	0.8	0.5270	0.8499	0.620	10.6128	0.2
0.9	0.4679	0.8838	0.5295	10.889	0.1	0.9	0.5284	0.8490	0.6224	10.6066	0.1
	cos	sin	cot	tan	deg		cos	sin	cot	tan	deg

deg	sin	cos	tan	cot		deg	sin	cos	tan	cot	
32.0	0.5299	0.8480	0.6249	10.603	58.0	36.0	0.5878	0.8090	0.7265	10.3764	54.0
0.1	0.5314	0.8471	0.6273	10.5941	0.9	0.1	0.5892	0.8080	0.7292	10.3713	0.9
0.2	0.5329	0.8462	0.6297	10.5880	0.8	0.2	0.5906	0.8070	0.7319	10.3663	0.8
0.3	0.5344	0.8453	0.6322	10.5818	0.7	0.3	0.5920	0.8059	0.7346	10.3613	0.7
0.4	0.5358	0.8443	0.6346	10.5757	0.6	0.4	0.5934	0.8049	0.7373	10.3564	0.6
0.5	0.5373	0.8434	0.6371	10.5697	0.5	0.5	0.5948	0.8039	0.740	10.3514	0.5
0.6	0.5388	0.8425	0.6395	10.5637	0.4	0.6	0.5962	0.8028	0.7427	10.3465	0.4
0.7	0.5402	0.8415	0.6420	10.5577	0.3	0.7	0.5976	0.8018	0.7454	10.3416	0.3
0.8	0.5417	0.8406	0.6445	10.5517	0.2	0.8	0.5990	0.807	0.7481	10.3367	0.2
0.9	0.5432	0.8396	0.6469	10.5458	0.1	0.9	0.604	0.7997	0.7508	10.3319	0.1
33.0	0.5446	0.8387	0.6494	10.5399	57.0	37.0	0.6018	0.7986	0.7536	10.3270	53.0
0.1	0.5461	0.8377	0.6519	10.5340	0.9	0.1	0.6032	0.7976	0.7563	10.3222	0.9
0.2	0.5476	0.8368	0.6544	10.5282	0.8	0.2	0.6046	0.7965	0.7590	10.3175	0.8
0.3	0.5490	0.8358	0.6569	10.5224	0.7	0.3	0.6060	0.7955	0.7518	10.3127	0.7
0.4	0.5505	0.8348	0.6594	10.5166	0.6	0.4	0.6074	0.7944	0.7646	10.3079	0.6
0.5	0.5519	0.8339	0.6619	10.5108	0.5	0.5	0.6088	0.7934	0.7673	10.3032	0.5
0.6	0.5534	0.8329	0.6644	10.5051	0.4	0.6	0.6101	0.7923	0.7701	10.2985	0.4
0.7	0.5548	0.8320	0.6669	10.4994	0.3	0.7	0.6115	0.7912	0.7729	10.2938	0.3
0.8	0.5563	0.8310	0.6694	10.4938	0.2	0.8	0.6129	0.7902	0.7757	10.2892	0.2
0.9	0.5577	0.830	0.6720	10.4882	0.1	0.9	0.6143	0.7891	0.7785	10.2846	0.1
34.0	0.5592	0.8290	0.6745	10.4826	56.0	38.0	0.6157	0.7880	0.7813	10.2799	52.0
0.1	0.5606	0.8281	0.6771	10.4770	0.9	0.1	0.6170	0.7869	0.7841	10.2753	0.9
0.2	0.5621	0.8271	0.6796	10.4715	0.8	0.2	0.6184	0.7859	0.7869	10.2708	0.8
0.3	0.5635	0.8261	0.6822	10.4659	0.7	0.3	0.6198	0.7848	0.7898	10.2662	0.7
0.4	0.5650	0.8251	0.6847	10.4605	0.6	0.4	0.6211	0.7837	0.7926	10.2617	0.6
0.5	0.5664	0.8241	0.6873	10.4550	0.5	0.5	0.6225	0.7826	0.7954	10.2572	0.5
0.6	0.5678	0.8231	0.6899	10.4496	0.4	0.6	0.6239	0.7815	0.7983	10.2527	0.4
0.7	0.5693	0.8221	0.6924	10.4442	0.3	0.7	0.6252	0.7804	0.8012	10.2482	0.3
0.8	0.5707	0.8211	0.6950	10.4388	0.2	0.8	0.6266	0.7793	0.8040	10.2437	0.2
0.9	0.5721	0.8202	0.6970	10.4335	0.1	0.9	0.6280	0.7782	0.8069	10.2393	0.1
35.0	0.5736	0.8192	0.702	10.4281	55.0	39.0	0.6293	0.7771	0.8098	10.2349	51.0
0.1	0.5750	0.8181	0.7028	10.4229	0.9	0.1	0.6307	0.7760	0.8127	10.2305	0.9
0.2	0.5764	0.8171	0.7054	10.4176	0.8	0.2	0.6320	0.7749	0.8156	10.2261	0.8
0.3	0.5779	0.8161	0.7080	10.4124	0.7	0.3	0.6334	0.7738	0.8185	10.2218	0.7
0.4	0.5793	0.8151	0.7107	10.4071	0.6	0.4	0.6347	0.7727	0.8214	10.2174	0.6
0.5	0.5807	0.8141	0.7133	10.4019	0.5	0.5	0.6361	0.7716	0.8243	10.2131	0.5
0.6	0.5821	0.8131	0.7159	10.3968	0.4	0.6	0.6474	0.7705	0.8273	10.2088	0.4
0.7	0.5835	0.8121	0.7186	10.3916	0.3	0.7	0.6388	0.7694	0.8302	10.2045	0.3
0.8	0.5850	0.8111	0.7212	10.3865	0.2	0.8	0.6401	0.7683	0.8332	10.202	0.2
0.9	0.5864	0.810	0.7239	10.3814	0.1	0.9	0.6414	0.7672	0.8361	10.1960	0.1
	cos	sin	cot	tan	deg		cos	sin	cot	tan	deg

deg	sin	cos	tan	cot		deg	sin	cos	tan	cot	
40.0	0.6428	0.7660	0.8291	10.1918	50.0	43.0	0.6820	0.7314	0.9325	10.724	47.0
0.1	0.6441	0.7649	0.8421	10.1875	0.9	0.1	0.6833	0.7302	0.9358	10.686	0.9
0.2	0.6455	0.7638	0.8451	10.1833	0.8	0.2	0.6845	0.7290	0.9391	10.649	0.8
0.3	0.6468	0.7627	0.8481	10.1792	0.7	0.3	0.6858	0.7278	0.9424	10.612	0.7
0.4	0.6481	0.7615	0.8511	10.1750	0.6	0.4	0.6871	0.7266	0.9457	10.575	0.6
0.5	0.6494	0.7604	0.8541	10.1708	0.5	0.5	0.6884	0.7254	0.9490	10.538	0.5
0.6	0.6508	0.7593	0.8571	10.1667	0.4	0.6	0.6896	0.7242	0.9523	10.501	0.4
0.7	0.6521	0.7581	0.8601	10.1626	0.3	0.7	0.6909	0.7230	0.9556	10.464	0.3
0.8	0.6534	0.7570	0.8632	10.1585	0.2	0.8	0.6921	0.7218	0.9590	10.428	0.2
0.9	0.6547	0.7559	0.8662	10.1544	0.1	0.9	0.6934	0.7206	0.9623	10.392	0.1
41.0	0.6561	0.7547	0.8693	10.1504	49.0	44.0	0.6947	0.7193	0.9657	10.355	46.0
0.1	0.6574	0.7536	0.8724	10.1463	0.9	0.1	0.6959	0.7181	0.9691	10.319	0.9
0.2	0.6587	0.7524	0.8754	10.1423	0.8	0.2	0.6972	0.7169	0.9725	10.283	0.8
0.3	0.660	0.7513	0.8785	10.1383	0.7	0.3	0.6984	0.7157	0.9759	10.247	0.7
0.4	0.6613	0.7501	0.8816	10.1343	0.6	0.4	0.6997	0.7145	0.9793	10.212	0.6
0.5	0.6626	0.7490	0.8847	10.1303	0.5	0.5	0.709	0.7133	0.9827	10.176	0.5
0.6	0.6639	0.7478	0.8878	10.1263	0.4	0.6	0.7022	0.7120	0.9861	10.141	0.4
0.7	0.6652	0.7466	0.8910	10.1224	0.3	0.7	0.7034	0.7108	0.9896	10.105	0.3
0.8	0.6665	0.7455	0.8941	10.1184	0.2	0.8	0.6794	0.7337	0.9260	10.799	0.2
0.9	0.6678	0.7443	0.8972	10.1145	0.1	0.9	0.6807	0.7325	0.9293	10.761	0.1
42.0	0.6691	0.7431	0.904	10.1106	48.0	45.0	0.7071	0.7071	1.0000	1.0000	45.0
0.1	0.6704	0.7420	0.9036	10.1067	0.9						
0.2	0.6717	0.7408	0.9067	10.1028	0.8						
0.3	0.6730	0.7396	0.9099	10.990	0.7						
0.4	0.6743	0.7385	0.9131	10.951	0.6						
0.5	0.6756	0.7373	0.9163	10.913	0.5						
0.6	0.6769	0.7361	0.9195	10.875	0.4						
0.7	0.6782	0.7349	0.9228	10.837	0.3						
0.8	0.6794	0.7337	0.9260	10.799	0.2						
0.9	0.6807	0.7325	0.9293	10.761	0.1						
	cos	sin	cot	tan	deg		cos	sin	cot	tan	deg

MODULE 2 INDEX

A

A.c. and d.c., alternating current electricity, 1-1
A.c. formulas, useful, AIV-l to AIV-2
A.c. generation, basic, 1-8 to 1-12
    cycle, 1-8 to 1-12
    frequency, 1-11
    period, 1-11
    wavelength, 1-12
Alternating current and capacitors, 4-6 to 4-9
Alternating current and inductance, 4-2 to 4-4
Alternating current values, 1-13 to 1-20
    average value, 1-14
    effective value of a sine wave, 1-15
    instantaneous value, 1-14
    peak and peak-to-peak values, 1-13
    sine waves in phase, 1-18
    sine waves out of phase, 1-19 to 1-20
Apparent power in a.c. circuits, calculating, 4-21, 4-22
Audiofrequency transformers, 5-22
Autotransformers, 5- 21
Average value, 1-14

C

Capacitance, 3-2 to 3-34
    capacitor losses, 3-7 to 3-11
        charging, 3-8 to 3-11
        discharging, 3-11
    capacitors in series and parallel, 3-20 to 3-26
        color codes, 3-27 to 3-33
        fixed capacitor, 3-22 to 3-26
        in parallel, 3-21
        in series, 3-20
        variable capacitor, 3-26
    charge and discharge of an RC series circuit, 3-11 to 3-15
        charge cycle, 3-12 to 3-13
        discharge cycle, 3-14 to 3-15
    electrostatic field, 3-2, 3-3
        factors affecting the value of capacitance, 3-5 to 3-7
        farad, 3-4

Capacitance—Continued
        simple capacitor, 3-3
    RC time constant, 3-16
    summary, 3-33 to 3-38
    universal time constant chart, 3-16 to 3-19
    voltage rating of capacitors, 3-7
Capacitive reactance, 4-8
Capacitor losses, 3-7 to 3-11
    charging, 3-8 to 3-11
    discharging, 3-11
Capacitors in series and parallel, 3-20 to 3-22
    color codes, 3-27 to 3-34
    fixed capacitor, 3-22 to 3-26
    in parallel, 3-21
    in series, 3-20
    variable capacitor, 3-26
Coefficient of coupling, transformers, 5-10
Coil, magnetic field, 1-6 to 1-8
    losses, 1-8
    polarity, 1-7
    strength, 1-8
Color codes for capacitors, 3-27 to 3-34
Concepts of alternating current electricity, 1-1 to 1-32
    a.c. and d.c., 1-1
    alternating current values, 1-13 to 1-20
    basic a.c. generation, 1-9 to 1-14
        cycle, 1-8 to 1-10
        frequency, 1-11
        period, 1-11
        wavelength, 1-12
    disadvantages of d.c. compared to a.c., 1-2
    electromagnetism, 1-4
    magnetic fields, 1-4 to 1-8
    around a current-carrying conductor, 1-4 to 1-6
    of a coil, 1-6 to 1-8
    Ohm's law in a.c. circuits, 1-21
    summary, 1-22 to 1-29
    voltage waveforms, 1-3
Copper loss, transformer, 5-18
Core characteristics, transformers, 5-3 to 5-4
    hollow-core, 5-4
    shell-core, 5-5

Counter emf, producing, transformers, 5-8
Current-carrying conductor, magnetic fields, 1-4 to 1-6
Cycle, a.c. generation, 1-8 to 1-11

D

Disadvantages of d.c. compared to a.c., 1-2

E

Eddy-current loss, transformer, 5-18
Effective value of a sine wave, 1-15 to 1-17
Effects of current on the body, 5-22
Efficiency, transformer, 5-18
Electric shock, 5-23 to 5-24
    preventing electric shock, 5-24 to 5-25
    safety, 5-22
Electromagnetism, 1-4
Electromotive force (emf), 2-2
Electrostatic field, 3-2 to 3-7
    factors affecting the value of capacitance, 3-5 to 3-7
    farad, 3-4
    simple capacitor, 3-3

F

Farad, capacitance, 3-5
Fixed capacitor, 3-22 to 3-26
Frequency, a.c. generation, 1-11

G

Glossary, AI-1 to AI-5
Greek alphabet, AII-1
Growth and decay of current in an LR series circuit, 2-10 to 2-14
    L/R time constant, 2-13 to 2-14

H

Hollow-core transformers, 5-4
Hysteresis loss, transformer, 5-18

I

Impedance, a.c. circuits, 4-11 to 4-15
Impedance-matching transformers, 5-22
Inducing a voltage in the secondary, transformers, 5-9
Inductance, 2-1 to 2-23
    characteristics, 2-2 to 2-19
        electromotive force (emf), 2-2
        factors affecting mutual inductance, 2-16
        growth and decay of current in an LR series circuit, 2-10 to 2-14
        mutual inductance, 2-15
        parallel inductors without coupling, 2-19
        power loss in an inductor, 2-14
        self-inductance, 2-4 to 2-12
        series inductors with magnetic coupling, 2-17
        series inductors without magnetic coupling, 2-17
        unit of inductance, 2-10
    summary, 2-19 to 2-25
Inductive and capacitive reactance, 4-2 to 4-34
    capacitors and alternating current, 4-6 to 4-8
        capacitive reactance, 4-8
    inductance and alternating current, 4-2 to 4-4
        inductive reactance, 4-5
    parallel RLC circuits, 4-28 to 4-34
    reactance, impedance, and power relationships in a.c. circuits, 4-9 to 4-25
        impedance, 4-11 to 4-15
        Ohms law for a.c., 4-15
        power in a.c. circuits, 4-17 to 4-22
        reactance, 4-9
    series RLC circuits 4-25 to 4-28
    summary, 4-34 to 4-38
Inductive reactance, 4-5
Instantaneous value, 1-14

L

Learning objectives, 1-1, 2-1, 3-1, 4-1, 5-1
    capacitance, 3-2
    concepts of alternating current electricity, 1-1

Learning objectives—Continued
    inductance, 2-2
    inductive and capacitive reactance, 4-1
    transformers, 5-1
Load, effect of, transformers, 5-13
Losses, transformer, 5-17 to 5-18
    copper loss, 5-18
    eddy-current loss, 5-18
    hysteresis loss, 5-18
L/R time constant, 2-13 to 2-14

M

Magnetic fields, 1-4 to 1-8
    around a current-carrying conductor, of a coil, 1-6 to 1-8
Mutual flux, transformers, 5-14
Mutual inductance, 2-15

N

Navy Electricity and Electronics Training Series, iv-vi
No-load condition, transformer, 5-8

O

Ohm's law for a.c., 4-15
Ohm's law in a.c. circuits, 1-21, 1-22

P

Parallel inductors without coupling, 2-19
Parallel RLC circuits, 4-28 to 4-34
Peak and peak-to-peak values, 1-13 to 1-15
Period, a.c. generation, 1-11
Power factor, a.c. circuits, 4-23
Power factor correction, a.c. circuits, 4-24
Power in a.c. circuits, 4-17 to 4-22
    calculating apparent power in a.c. circuits, 4-21, 4-22
    calculating reactive power in a.c. circuits, 4-20
    calculating true power in a.c. circuits, 4-18
    power factor, 4-23
    power factor correction, 4-24
Power loss in an inductor, 2-14

Power relationship between primary and secondary windings, transformers, 5-17
Power transformers, 5-20
Primary and secondary phase relationship, transformers, 5-9

R

Radiofrequency transformers, 5-22
Ratings, transformers, 5-19
RC series circuit, charge and discharge, 3-11 to 3-14
    charge cycle, 3-8 to 3-11
    discharge cycle, 3-11
RC time constant, 3-16
Reactance, a.c. circuits, 4-9
Reactive power in a.c. circuits, calculating, 4-20
RLC circuits, 4-25 to 4-28, 4-28 to 4-34
    parallel, 4-28 to 4-34
    series, 4-25 to 4-28

S

Safety, transformers, 5-22 to 5-24
Schematic symbols for transformers, 5-7
Self-inductance, 2-4 to 2-13
    factors affecting coil inductance, 2-7 to 2-12
Series inductors with magnetic coupling, 2-17
Series inductors without magnetic coupling, 2-17
Series RLC circuits, 4-25 to 4-28
Shell-core transformers, 5-5
Shock victims, rescue and care, 5-23
Simple capacitor, 3-3
Sine waves in phase, 1-18
Sine waves out of phase, 1-19 to 1-20
Square and square roots, AIII-1

T

Transformers, 5-1 to 5-28
    basic operation, 5-2
    components, 5-3 to 5-5
        core characteristics, 5-3 to 5-5
        transformer windings, 5-5 to 5-6

Transformers—Continued
- how a transformer works, 5-8 to 5-19
  - coefficient of coupling, 5-10
  - effect of a load, 5-13
  - inducing a voltage in the secondary, 5-9
  - mutual flux, 5-14
  - no-load condition, 5-8
  - power relationship between primary and secondary windings, 5-17
  - primary and secondary phase relationship, 5-9
  - producing a counter emf, 5-8
  - transformer efficiency, 5-18 to 5-19
  - transformer losses, 5-17
  - turns and current ratios, 5-14 to 5-16
  - turns and voltage ratios, 5-11
- ratings, 5-19
- safety, 5-22 to 5-24
  - effects of current on the body, 5-22
  - electric shock, 5-23
  - precautions for preventing electric shock, 5-24 to 5-25
  - rescue and care of shock victims, 5-23
- schematic symbols for transformers, 5-7
- summary, 5-25 to 5-28
- types and applications, 5-20 to 5-22
  - audiofrequency transformers, 5-22

Transformers—Continued
- autotransformers, 5-21
- impedance-matching transformers, 5-22
- power transformers, 5-20
- radiofrequency transformers, 5-22

Trigonometric functions, AV-1 to AV-4
Trigonometric tables, AVI-1 to AVI-7
True power in a.c. circuits, calculating, 4-18
Turns and current ratios, transformers, 5-14 to 5-16
Turns and voltage ratios, transformers, 5-11 to 5-13

U

Unit of inductance, 2-10
Universal time constant chart, 3-16 to 3-19

V

Variable capacitor, 3-26
Voltage rating of capacitors, 3-7
Voltage waveforms, 1-3

W

Wavelength, a.c. generation, 1-12
Windings transformer, 5-5 to 5-6

Assignment Questions

Information: The text pages that you are to study are provided at the beginning of the assignment questions.

ASSIGNMENT 1

Textbook assignment: Chapter 1, "Concepts of Alternating Current," pages 1-1 through 1-33.

1-1. Alternating current can be defined as current that varies in

1. amplitude and direction
2. magnitude and phase
3. amplitude and time
4. time and phase

1-2. Before a 120-volt dc source can be used to power a 12-volt load, the voltage must be reduced. Which of the following methods can be used?

1. A resistor placed in parallel with the load
2. A resistor placed in series with the load
3. A step-down transformer placed in series with load
4. A step-down transformer placed in parallel-with the load

1-3. Alternating current has replaced direct current in modern transmission systems because it has which of the following advantages?

1. Ac can be transmitted with no line loss
2. Ac can be transmitted at higher current levels
3. Ac can be transmitted at lower voltage levels
4. Ac can be readily stepped up or down

1-4. A waveform is a graphic plot of what quantities?

1. Current versus time
2. Amplitude versus time
3. Voltage versus amplitude
4. Magnitude versus amplitude

1-5. Which of the following properties surrounds a current-carrying conductor?

1. A magnetic field
2. A repulsive force
3. An attractive force
4. An electrostatic field

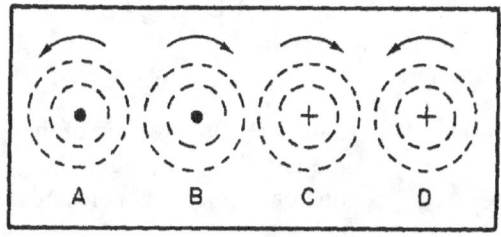

Figure 1A.—Conductors, cross-sectional view.

IN ANSWERING QUESTIONS 1-6 AND 1-7, REFER TO FIGURE 1A.

1-6. The direction of the magnetic field is correctly depicted by which of the followings

1. A and B
2. B and D
3. A and C
4. B and C

IN ANSWERING QUESTION 1-7, REFER TO FIGURE 1A AND IGNORE THE MAGNETIC FIELD ARROWS SHOWN IN THE FIGURE.

1-7. In which conductors will the magnetic fields (a) aid, and (b) oppose each other?

1. (a) A and C, (b) C and D
2. (a) A and D, (b) B and C
3. (a) A and C, (b) B and D
4. (a) A and B, (b) A and D

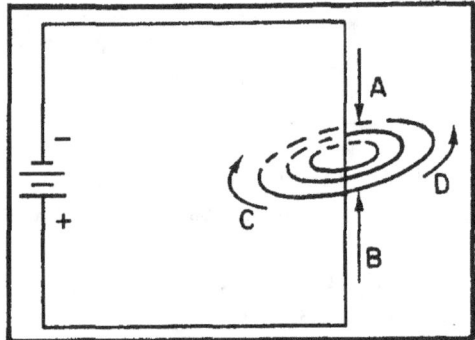

Figure 1B.—A magnetic field surrounding a current-carrying conductor.

IN ANSWERING QUESTIONS 1-8, REFER TO FIGURE 1B.

1-8. In figure 1B, the direction of the magnetic field surrounding the conductor is correctly indicated by what arrow?

1. A
2. B
3. C
4. D

1-9. Which of the following statements accurately describes the magnetic field surrounding a current-carrying conductor?

1. It is parallel to and equal along all parts of the conductor
2. It is parallel to and maximum at the most negative part of the conductor
3. It is perpendicular to and equal along all parts of the conductor
4. It is perpendicular to the conductor and maximum at the most negative point of the conductor

1-10. Which of the following factors determine(s) the intensity of a magnetic field surrounding a coil?

1. The amount of current flow through the coil
2. The type of core material
3. The number of turns in the conductor
4. All of the above

1-11. When you grasp a coil in your left hand with your thumb pointing in the direction of the north pole, your fingers will be wrapped around the coil in the direction of the

1. voltage potential
2. magnetic field
3. current flow
4. south pole

1-12. The power consumed in a conductor in realigning the atoms which set up the magnetic field is known as what type of loss?

1. Hysteresis loss
2. Magnetic loss
3. Field loss
4. Heat loss

1-13. The magnetic field surrounding a straight conductor is (a) what shape, and (b) is in what position relative to the conductor?

1. (a) Linked oblong
   (b) Parallel
2. (a) Concentric circles
   (b) Parallel
3. (a) Linked oblong
   (b) Perpendicular
4. (a) Concentric circles
   (b) Perpendicular

1-14. Why is a two-pole magnetic field set up around a coil?

1. Because separate lines of magnetic force link and combine their effects
2. Because concentric lines of force cross at right angles and combine.
3. Because lines of force are separated and bent at the coil ends
4. Because separate lines of force are attracted to the two poles of the coi

1-15. When a conductor is moving parallel to magnetic lines of force, (a) what relative number of magnetic lines are cut, and (b) what relative value of emf is induced?

1. (a) Minimum, (b) maximum
2. (a) Minimum, (b) minimum
3. (a) Maximum, (b) maximum
4. (a) Maximum, (b) minimum

1-16. When the induced voltage in a conductor rotating in a magnetic field is plotted against the degrees of rotation, the plot will take what shape?

1. A circle
2. A sine curve
3. A square wave
4. A straight line

1-17. When a loop of wire is rotated through 360° in a magnetic field, the induced voltage will be zero at which of the following points?

1. 45°
2. 90°
3. 180°
4. 270°

1-18. When a loop of wire is rotated 360° in a magnetic field, at what points will the induced voltage reach its maximum (a) positive, and (b) negative values?

1. (a) 0°, (b) 180°
2. (a) 0°, (b) 270°
3. (a) 90°, (b) 180°
4. (a) 90°, (b) 270°

1-19. When a coil of wire makes eight complete revolutions through a single magnetic field, (a) what total number of alternations of voltage will be generated and, (b) what total number of cycles of ac will be generated?

1. (a) 32, (b) 16
2. (a) 16, (b) 8
3. (a) 8, (b) 4
4. (a) 4, (b) 2

1-20. According to the left-hand rule for generators, when your thumb points in the direction of rotation, your (a) forefinger and (b) your middle finger will indicate the relative directions of what quantities?

1. (a) Current,
   (b) Magnetic flux, south to north
2. (a) Current,
   (b) Magnetic flux, north to south
3. (a) Magnetic flux, south to north,
   (b) Current
4. (a) Magnetic flux, north to south,
   (b) Current

1-21. Continuous rotation of a conductor through magnetic lines of force will produce what type of (a) voltage and (b) waveform?

1. (a) Ac, (b) sine wave
2. (a) Dc, (b) continuous level
3. (a) Ac, (b) sawtooth
4. (a) Dc, (b) pulsating wave

1-22. What is the term for the number of complete cycles of ac produced in one second?

1. Period
2. Waveform
3. Frequency
4. Wavelength

1-23. What is the unit of measurement for frequency?

1. Cycle
2. Hertz
3. Period
4. Maxwell

1-24. A loop of wire rotating at 60 rpm in a magnetic field will produce an ac voltage of what frequency?

1. 1 Hz
2. 60 Hz
3. 120 Hz
4. 360 Hz

1-25. An ac voltage of 250 hertz has a period of

1. 0.004 second
2. 0.025 second
3. 0.4 second
4. 2.5 seconds

1-26. What is the approximate frequency of an ac voltage that has a period of .0006 second?

1. 6 Hz
2. 16.67 Hz
3. 600 Hz
4. 1667 Hz

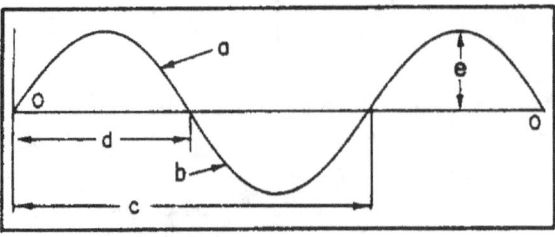

Figure 1C.—Components of a sinewave.

IN ANSWERING QUESTIONS 1-27 THROUGH 1-36, REFER TO FIGURE 1C. IN ANSWERING QUESTIONS 1-27 THROUGH 1-31, SELECT FROM COLUMN B THE COMPONENT THAT IS DESCRIBED BY THE TERM IN COLUMN A.

A. TERM	B. COMPONENT
1-27. Period	1. a
1-28. Negative alternation	2. b
	3. c
1-29. Wavelength	4. d
1-30. One-half cycle	
1-31 Postive alternation	

1-32. Component a is a measure of what quantity?

1. Frequency
2. Polarity
3. Amplitude
4. Time

1-33. Component a differs from component b in which of the following characteristics

1. Frequency
2. Polarity
3. Amplitude
4. Period

1-34. Component c would represent what quantities if it were expressed as (a) physical distance, and (b) time?

1. (a) Frequency   (b) period
2. (a) Period   (b) wavelength
3. (a) Frequency   (b) wavelength
4 (a) Wavelength   (b) period

1-35. The combined values of components a and b represent what ac value?

1. Peak-to-peak value
2. Average value
3. Effective value
4. Instantaneous value

1-36. A peak voltage is represented by which of the following components?

1. a
2. c
3. d
4. e

1-37. An ac voltage has a frequency of 350 Hz. In two seconds, what total number of times will the peak value of voltage be generated?

1. 350 times
2. 700 times
3. 1400 times
4. 2800 times

1-38. The value of current of an ac waveform taken at any particular moment of time is what type of value?

1. Average value
2. Effective value
3. Instantaneous value
4. Peak-to-peak value

1-39. While the value of an ac voltage may be expressed as one of several values, the accepted practice is to express it as what type value?

1. Average value
2. Instantaneous value
3. Peak-to-peak value
4. Effective value

1-40. The total of ten instantaneous values of an alternation divided by ten is equal to what value?

1. The peak value
2. The average value
3. The instantaneous value
4. The effective value

1-41. Which of the following mathematical formulas is used to find the average value of voltage for an ac voltage?

1. $E_{avg} = 0.707 \times E_{max}$
2. $E_{avg} = 1.414 \times E_{eff}$
3. $E_{avg} = 0.636 \times E_{max}$
4. $E_{avg} = 0.226 \times E_{eff}$

1-42. What is the average value of all of the instantaneous voltages occurring during one cycle of an ac waveform with a peak value of 60 volts?

1. 0 volts
2. 38 volts
3. 76 volts
4. 128 volts

1-43. If an ac voltage has an E_{max} of 220 volts, what is E_{avg}?

1. 50 volts
2. 140 volts
3. 156 volts
4. 311 volts

1-44. If an ac waveform has a peak-to-peak value of 28 volts, what is E_{avg}?

1. 40 volts
2. 20 volts
3. 18 volts
4. 9 volts

1-45. If an ac waveform has a peak value of 4.5 amperes, what is its average value?

1. 2.9 amperes
2. 3.2 amperes
3. 5.7 amperes
4. 6.4 amperes

1-46. If the average value of current of an ac waveform is 1.2 amperes, what is its maximum value of current?

1. 0.8 amperes
2. 0.9 amperes
3. 1.7 amperes
4. 1.9 amperes

1-47. The value of alternating current that will heat a resistor to the same temperature as an equal value of direct current is known as

1. I_{avg}
2. I_{eff}
3. I_{in}
4. I_{max}

1-48. The rms value for an ac voltage is equal to what other ac value?

1. E_{avg}
2. E_{max}
3. E_{eff}
4. E_{in}

1-49. What value will result by squaring all values for E_{inst}, averaging these values, and then taking the square root of that average?

1. E_{avg}
2. E_{max}
3. E_{eff}
4. E_{in}

1-50. The accepted, nominal value for household power in the United States is 120-volts, 60 Hz. What is the value of maximum voltage?

1. 170 volts
2. 120 volts
3. 85 volts
4. 76 volts

1-51. An ac voltmeter is usually calibrated to read which of the following ac values?

1. Average
2. Effective
3. Peak
4. Peak-to-peak

1-52. If the maximum value for an ac voltage is known, the E_{eff} can be found by using which of the following formulas?

1. $E_{eff} = E_{max}/.636$
2. $E_{eff} = E_{max}/.707$
3. $E_{eff} = E_{max} \times .707$
4. $E_{eff} = E_{max} \times 1.414$

1-53. If the I_{eff} of an ac waveform is 3.25 amperes, what is I_{max}?

1. 4.6 amperes
2. 2.3 amperes
3. 2.1 amperes
4. 1.6 amperes

1-54. If the rms value of the voltage of an ac waveform is 12.4 volt what is its average value? (Hint: compute E_{max} first.)

1. 8 volts
2. 11 volts
3. 15 volts
4. 18 volts

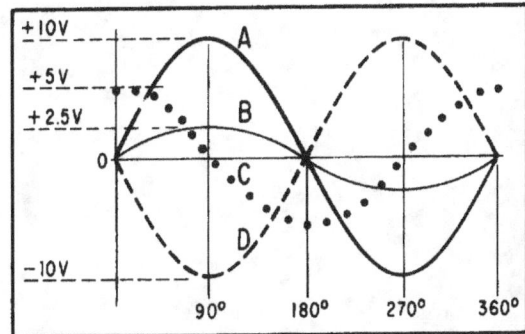

Figure 1D.—Phase relationship of sinewaves.

IN ANSWERING QUESTIONS 1-55 THROUGH 1-60, REFER TO FIGURE 1D.

1-55. What two waveforms are in phase?

1. A and B
2. A and C
3. C and D
4. B and C

1-56. What is the phase difference, if any, between waveform B and C?

1. B is 225° out of phase with C
2. B is 180° out of phase with C
3. B is 90° out of phase with C
4. None; they are in phase

1-57. What is the phase difference, if any, between waveform A and D?

1. A is 270° out of phase with D
2. A is 180° out of phase with D
3. A is 90° out of phase with D
4. None; they are in phase

1-58. If the voltage represented by waveform A is summed to the voltage represented by waveform D, what is the resultant voltage?

1. 20 volts
2. 15 volts
3. 10 volts
4. 0 volts

1-59. What is E_{in} at 90° that results from adding waveform B to waveform D?

1. +7.5 volts
2. +2.5 volts
3. -7.5 volts
4. -10 volts

1-60. What is the phase difference between waveform A and waveform C?

1. A lags C by 90°
2. A leads C by 90°
3. A leads C by 180°
4. A lags C by 180°

1-61. Which of the following is an important rule to remember when using Ohm's Law to solve ac circuit problems?

1. Always solve for resistance first
2. Give the answer as effective value
3. Never mix values
4. Convert all given values to effective before attempting to solve

1-62. An ac circuit is composed of three 20-ohm resistors connected in parallel. The average voltage supplied to this circuit is 62-volts ac. What is the maximum current?

1. 9.3 amperes
2. 14.6 amperes
3. 17.5 amperes
4. 22.5 amperes

1-63. If the ac source in question 1-62 is raised to an average value of 120 volts, what is the I_{eff}?

1. 11.48 amperes
2. 12.70 amperes
3. 20.01 amperes
4. 25.52 amperes

1-64. If E_{eff} is 150 volts and I_{max} is 4.5 amperes, what is the total resistance (R_T) of a circuit?

1. 21.2Ω
2. 23.6Ω
3. 33.3Ω
4. 47.1Ω

ASSIGNMENT 2

Textbook assignment: Chapter 2, "Inductance," pages 2-1 through 2-27.

2-1. The property of inductance offers opposition to which of the following quantities?

1. Constant current
2. Constant voltage
3. Changes in current
4. Changes in voltage

2-2. What is the symbol for inductance?

1. L
2. H
3. X_L
4. IND

2-3. What is the unit of measurement for inductance?

1. Ohm
2. Rel
3. Farad
4. Henry

2-4. If 9 volts are induced in a conductor when the current changes by 4.5 amperes in one second, what is the total inductance of the circuit?

1. 1.5 henries
2. 2.0 henries
3. 13.5 henries
4. 40.0 henries

2-5. What physical property is similar to inductance?

1. Mass
2. Motion
3. Velocity
4. Inertia

2-6. The difference in potential across a resistor, created by current through the resistor is an example of which of the following forces?

1. Resistive
2. Inertia
3. Inductive
4. Electromotive

2-7. When a magnetic field moves through a stationary conductor, the electrons in orbit are affected in what manner?

1. They are dislodged from orbit
2. They move closer to their nucleus
3. They move closer to other orbiting electrons
4. They bunch up on one side of the nucleus

2-8. When electrons are moved in a conductor by a magnetic field, a force known by which of the following terms is created?

1. Voltage
2. Electromotive
3. Potential difference
4. All of the above

2-9. Self-induced emf is also known as what force?

1. magnetic force
2. Inertial force
3. Electromotive force
4. Counter electromotive force

9

2-10. According to Lenz's Law, the induced emf produced by a change in current in an inductive circuit tends to have what effect on the current?

1. It aids a rise in current and opposes fall in current
2. It aids a fall in current and opposes a rise in current
3. It opposes either a rise or a fall in current
4. It aids either a rise or fall in current

2-11. The direction of the induced voltage in an inductor may be found by application of which of the following rules?

1. The left-hand rule for inductors
2. The left-hand rule for generators
3. The right-hand rule for conductors
4. The right-hand rule for motors

2-12. The left-hand rule for generators states that the thumb of the left hand points in the direction of motion of the

1. conductor
2. magnetic field
3. generator poles
4. induced current

2-13. When source voltage is removed from a current-carrying conductor, a voltage will be induced in the conductor by which of the following actions?

1. The decreasing voltage
2. The collapsing magnetic field
3. The reversal of current
4. The reversing electrical field

2-14. The property of inductance is present in which of the following electrical circuits?

1. An ac circuit
2. A dc circuit
3. A resistive circuit
4. Each of the above

2-15. How are inductors classified?

1. By core type
2. By conductor type
3. By the number of turns
4. By the direction of the windings on the core

2-16. Normally, most coils have cores composed of either air or

1. copper
2. carbon
3. soft iron
4. carbon steel

2-17. The hollow form of nonmagnetic material found in the center of an air-core coil has what purpose?

1. To focus the magnetic flux
2. To support the windings
3. To act as a low resistance path for flux
4. To serve as a container for the core

2-18. Which of the following factors will NOT affect the value of inductance of a coil?

1. Number of coil turns
2. Diameter of the coil
3. Conductor tensility
4. Core materials used

2-19. When the number of turns is increased in a coil from 2 to 4, the total inductance will increase by a factor of

1. eight
2. two
3. six
4. four

2-20. Why do large diameter coils have greater inductance than smaller diameter coils, all other factors being the same?

1. Large diameter coils have more wire and thus more flux
2. Large diameter coils have less resistance
3. Small diameter coils have less resistance
4. Small diameter coils have large cemfs which oppose current flow

2-21. If the radius of a coil is doubled, its inductance is increased by what factor?

1. One
2. Two
3. Eight
4. Four

2-22. If the length of a coil is doubled while the number of turns is kept the same, this will have (a) what effect on inductance and (b) by what factor?

1. (a) Decrease, (b) by 1/4
2. (a) Decrease, (b) by 1/2
3. (a) Increase, (b) by 2 times
4. (a) Increase, (b) by 4 times

2-23. A soft iron core will increase inductance because it has which of the following characteristics?

1. Low permeability and low reluctance
2. Low permeability and high reluctance
3. High permeability and high reluctance
4. High permeability and low reluctance

2-24. An increase in the permeability of the core of a coil will increase which of the following coil characteristics?

1. Magnetic flux
2. Reluctance
3. Resistance
4. Conductance

2-25. If a coil is wound in layers, its inductance will be greater than that of a similar single-layer coil because of a higher

1. permeability
2. flux linkage
3. reluctance
4. conductance

2-26. Regardless of the method used, inductance of a coil can ONLY be increased by increasing what coil characteristic?

1. Transconductance
2. Reluctance
3. Flux linkage
4. Conductance

2-27. What is the symbol used to denote the basic unit of measurement for inductance?

1. L
2. H
3. I
4. F

2-28. What does the Greek letter Delta signify as in "ΔI" or "Δt"?

1. The values are constant
2. The values are average
3. The values are changing
4. The values are effective

2-29. An electrical circuit contains a coil. When the current varies 2.5 amperes in one second, 7.5 volts are induced in the coil. What is the value of inductance of the coil?

1. 1 henry
2. 2.2 henries
3. 3.3 henries
4. 4 henries

2-30. An ac electrical current varies 1.5 amperes in one second and is applied to a 10-henry coil. What is the value of the emf induced across the coil?

1. 1.0 volt
2. 1.5 volts
3. 11.5 volts
4. 15.0 volts

2-31. If a coil is rated at 10 henries, what is its value in (a) millihenries and (b) microhenries?

1. (a) 10,000 mH, (b) 10,000,000 μH
2. (a) 10,000 mH, (b) 1,000,000 μH
3. (a) 1,000 mH, (b) 1,000,000 μH
4. (a) 1,000 mH, (b)   100,000 μH

THIS SPACE LEFT BLANK INTENTIONALLY.

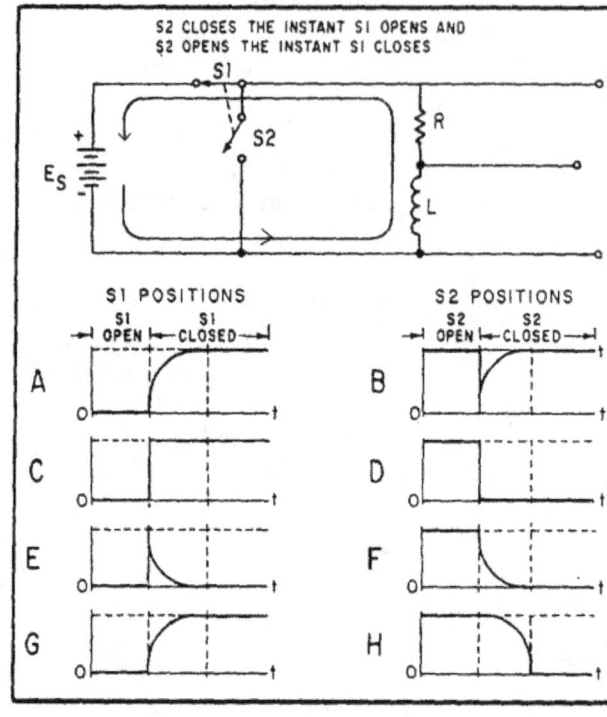

Figure 2A.—LR circuit characteristics.

IN ANSWERING QUESTIONS 2-32 THROUGH 2-40, REFER TO FIGURE 2A.

2-32. What waveform is an illustration of the voltage (E_S) present across the voltage divider when switch S1 is closed?

1. B
2. C
3. E
4. G

2-33. The voltage dropped across R when switch S1 is closed is depicted in which of the following waveforms?

1. G
2. H
3. F
4. C

2-34. The voltage developed across L when switch S1 is closed is depicted in what waveform?

1. A
2. B
3. E
4. H

2-35. Which of the following waveforms depicts growth current (I_g) through the coil (L)?

1. A
2. D
3. E
4. H

2-36. What waveform depicts the voltage developed across R when switch S2 is closed?

1. E
2. F
3. H
4. D

2-37. During the first instant when switch S1 is closed, maximum voltage is dropped across

1. the battery
2. the resistor
3. the coil
4. both the coil and resistor

2-38. During the first instant when switch S1 is closed, current is maximum in which, if any, of the following parts of the h circuit?

1. The battery
2. The coil
3. The resistor
4. None of the above

2-39. In the first instant when switch S1 is closed, the entire battery-voltage is used to overcome the

1. resistance of R
2. resistance of L
3. emf developed in R
4. emf developed in L

2-40. When switch S2 is closed, energy is supplied to the circuit by the

1. battery through S_2
2. battery through S_1
3. collapsing magnetic field of L_1
4. expanding magnetic field of L_1

2-41. One L/R time constant is equal to the time required for the current in an inductor to reach what portion of its maximum value?

1. 63.2%
2. 37.8%
3. 25.2%
4. 12.8%

2-42. Maximum current will flow in an LR circuit after a minimum of how many time constants have elapsed?

1. One
2. Five
3. Three
4. Four

2-43. The maximum current in an LR circuit is 20 amperes. What total current will be flowing in the circuit at the end of the second time constant of the charge cycle?

1. 20.0 amperes
2. 17.3 amperes
3. 12.6 amperes
4. amperes

2-44. Refer to the circuit described in question 2-43. Circuit current will increase by what amount during the second time constant?

1. 17.3 amperes
2. 12.6 amperes
3. 7.6 amperes
4. 4.7 amperes

2-45. An LR circuit has a maximum current of 30 mA. At the end of the first time constant of the discharge cycle, what total current will be flowing in the circuit?

1. 11 mA
2. 19 mA
3. 26 mA
4. 28 mA

2-46. An LR circuit contains a 150-ohm resistor and a 2-henry coil. What is the time value of one L/R time constant?

1. 7.5 seconds
2. .75 seconds
3. 1.33 seconds
4. .0133 seconds

2-47. An LR circuit has a time constant of .05 second and an inductor with a value of .60 henry. What value of resistor is required?

1. 5 ohms
2. 12 ohms
3. 24 ohms
4. 64 ohms

2-48. An LR circuit is composed of a coil of .5 henry and a 10-ohm resistor. The maximum current in the circuit is 5 amperes. After the circuit is energized, how long will it take for the current to reach maximum value?

1. 1.0 second
2. 0.05 second
3. 0.25 second
4. 5.0 seconds

2-49. Inductors experience copper loss for what reason?

1. Because of flux leakage in the copper core
2. Because the reactance of an inductor is greater than the resistance of an inductor
3. Because all inductors have resistance which dissipates power
4. Because the inertia of the magnetic field must be overcome every time the direction of current changes

2-50. Copper loss of an inductor can be calculated by the use of which of the following formulas?

1. $P = I^2 R$
2. $P = I^2 E$
3. $P = \dfrac{E}{R^2}$
4. $P = \dfrac{I^2}{R}$

2-51. What term applies to the power loss in an iron core inductor due to the current induced in the core?

1. Iron loss
2. Heat loss
3. Hysteresis loss
4. Eddy-current loss

2-52. Power consumed by an iron core inductor in reversing the magnetic field of the core is termed as what type of loss?

1. Iron loss
2. Heat loss
3. Hysteresis loss
4. Eddy-current loss

2-53. When does mutual inductance occur between inductors?

1. Whenever eddy-currents do not exist
2. Whenever the flux of one inductor causes an emf to be induced in another inductor
3. Whenever the effect of one inductor is aided by another inductor
4. Whenever the effect of one inductor is opposed by another inductor

2-54. Mutual inductance between two coils is affected by which of the following factors?

1. Material of the windings
2. Physical dimensions of the coils
3. Direction of the coil windings
4. Hysteresis characteristics of the coils

2-55. The coefficient of coupling between two coils is a measure of what factor?

1. The turns ratio of the coils
2. The distance between the coils
3. The relative positive of the coils
4. The magnetic flux ratio linking the coils

2-56. Two coils have a coefficient of coupling of .7 and are rated at 12 µH and 3 µH respectively. What is their total mutual inductance?

1. 4.2 µH
2. 5.2 µH
3. 7.0 µH
4. 10.5 µH

2-57. An electrical circuit contains four non-coupled inductors in a series configuration. The inductors have the following values: 2 µH, 3.5 µH, 6 µH, and 1 µH. What is the total inductance (L_T) of the circuit?

1. 45.0 µH
2. 42.5 µH
3. 12.5 µH
4. 11.5 µH

2-58. Two inductors of 3.6 µH and 7.3 µH are wired together in series and they aid each other. The mutual inductance for the circuit is 3.6µH. What is the total inductance (L_T) Of the circuit?

1. 17.5 µH
2. 18.1 µH
3. 24.8 µH
4. 34.4 µH

2-59. An electrical circuit contains three non-coupled inductors of 3.3 µH, 4.5 µH, and 2.0 µH wired in parallel. What is the total inductance of the circuit?

1. 9.8 µH
2. 3.6 µH
3. 0.98 µH
4. 0.28 µH

ASSIGNMENT 3

Textbook assignment: Chapter 3, "Capacitance," pages 3-1 through 3-41.

3-1. Capacitance and inductance in a circuit are similar in which of the following ways?

1. Both oppose current
2. Both aid voltage
3. Both cause the storage of energy
4. Both prevent the storage of energy

3-2. Capacitance is defined as the property of a circuit that

1. opposes a change in voltage
2. aids a change in voltage
3. opposes a change in current
4. aids a change in current

3-3. A capacitor is a device that stores energy in a/an

1. electrostatic field
2. electromagnetic field
3. induced field
4. molecular field

3-4. Electrostatic fields have what effect on (a) free electrons, and (b) bound electrons?

1. (a) Attracts them to the negative charges
   (b) Frees them from their orbits
2. (a) Attracts them to the positive charges
   (b) Frees them from their orbits
3. (a) Attracts them to the negative charges
   (b) Distorts their orbits
4. (a) Attracts them to the positive charges
   (b) Distorts their orbits

3-5. The influence of a charge on an electron orbit is correctly depicted by which of the following illustrations?

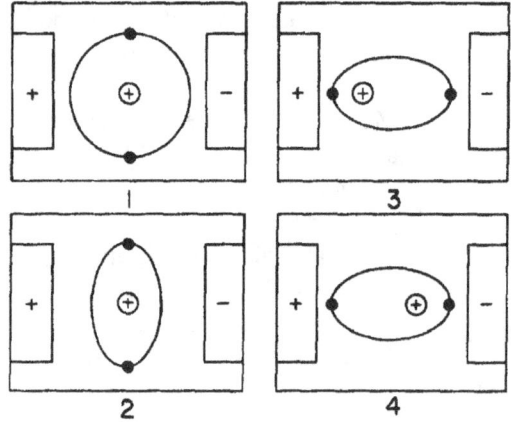

3-6. Electrostatic lines of force radiate from a charged particle along what type of lines?

1. Straight lines
2. Curved lines
3. Elliptical lines
4. Orbital lines

THIS SPACE LEFT BLANK INTENTIONALLY.

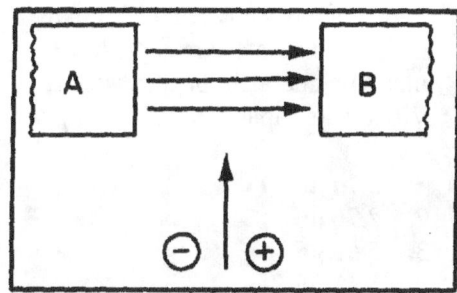

Figure 3A.—Electron and proton entering an electrostatic field.

IN ANSWERING QUESTION 3-7, REFER TO FIGURE 3A.

3-7. When the illustrated electron and proton enter the electrostatic field, toward what plate(s), will the (a) electron and, (b) proton be deflected?

1. (a) A (b) B
2. (a) B (b) A
3. (a) A (b) A
4. (a) B (b) B

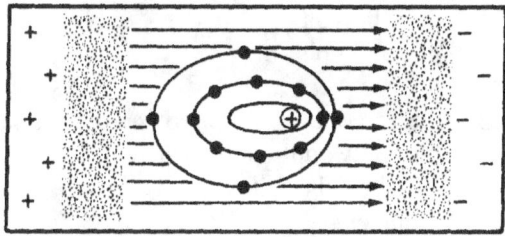

Figure 3B.—Effect of electrostatic lines of force.

IN ANSWERING QUESTION 3-8, REFER TO FIGURE 3B.

3-8. If the charges on the two plates are reversed, what will happen to the electrons?

1. They will dislodge from the atom
2. They will stay where they are
3. They will go back to circular orbits
4. They will distort in the opposite direction

3-9. Which of the following combinations describe(s) a simple capacitor?

1. Two copper plates separated by an iron plate
2. Two copper plates separated by a sheet of mica
3. Two iron plates separated by an air gap
4. Both 2 and 3 above

3-10. A capacitor that stores 6 coulombs of electrons when a potential of 2 volts is applied across its terminals has what total value of capacitance?

1. 12 farads
2. 8 farads
3. 3 farads
4. 6 farads

IN ANSWERING QUESTIONS 3-11 THROUGH 3-14, MATCH THE TERMS IN COLUMN B WITH THEIR MATHEMATICAL VALUES IN COLUMN A.

	A. VALUES		B. TERMS
3-11.	.000001 F	1.	Farad
3-12.	1×10^{-12} F	2.	Microfarad
3-13.	1×10^{-6} F	3.	Picofarad
3-14.	1×10^{0} F		

3-15. A capacitor of .0069 microfarad has which of the following capacitance values when measured in picofarads?

1. .000069 pF
2. 6900 pF
3. 6.9×10^{-9} pF
4. Both 2 and 3 above, individually, are correct

3-16. Which of the following characteristics of a capacitor can be varied WITHOUT altering its capacitance?

1. Area of the plates
2. Thickness of the dielectric
3. Material of the dielectric
4. Thickness of the plates

3-17. Which of the following actions will increase the capacitance of a capacitor?

1. The plates are moved closer together
2. The plates are moved farther apart
3. The dielectric constant is decreased
4. Both 2 and 3 above

3-18. Two capacitors are identical with the exception of the material used for the dielectric. Which of the following combinations of dielectric material will cause capacitor (b) to have a larger capacitance than capacitor (a)?

1. (a) Glass   (b) Paraffin paper
2. (a) Glycerine   (b) Pure water
3. (a) Petroleum   (b) Air
4. (a) Paraffin paper (b) Petroleum

3-19. Two capacitors are identical with the exception of the material used for the dielectric. Which of the following combinations of dielectric materials will cause the capacitors to have almost the same capacitance?

1. Glass, paraffin paper
2. Mica, petroleum
3. Vacuum, air
4. Petroleum, rubber

3-20. A capacitor is composed of two plates. Each plate has an area of 7 square inches. The plates are separated by a 2-inch thick paraffin paper dielectric. What is its capacitance?

1. 2.76 µF
2. 2.76 pF
3. 5.51 µF
4. 5.51 pF

3-21. The maximum voltage that can be applied to a capacitor without causing current flow through the dielectric is called

1. breaking voltage
2. limiting voltage
3. conduction voltage
4. working voltage

3-22. A capacitor with a working voltage of 300 volts would normally have what maximum effective voltage applied to it?

1. 200 volts
2. 250 volts
3. 300 volts
4. 350 volts

3-23. An ac voltage of 350 volts effective can be safely applied to a capacitor with which of the following working voltages?

1. 550 volts
2. 400 volts
3. 350 volts
4. 250 volts

3-24. Which, if any, of the following conditions may cause a capacitor to suffer power losses?

1. Dielectric hysteresis
2. Plate loading
3. Plate heating
4. None of the above

3-25. Rapid reversals in the polarity of the line voltage applied to a capacitor will cause what type of capacitor power loss?

1. Dielectric-leakage
2. Dielectric-hysteresis
3. Plate-loading
4. Plate-leakage

3-26. What type of dielectric is LEAST sensitive to power dielectric-hysteresis losses?

1. Pure water
2. Air
3. Vacuum
4. Mica

3-27. As the current through a capacitor increases, which of the following types of capacitor losses will increase?

1. Dielectric-hysteresis
2. Dielectric-leakage
3. Plate-leakage
4. Plate-breakdown

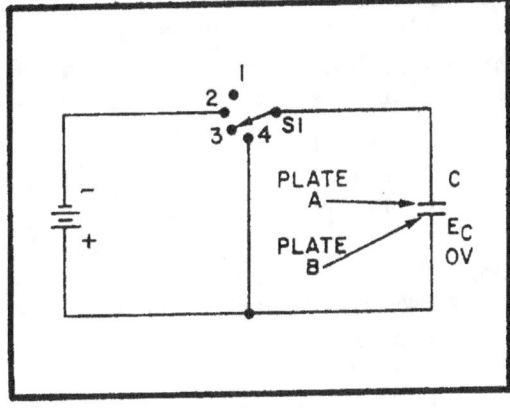

Figure 3C.—Simple capacitor circuit.

IN ANSWERING QUESTIONS 3-28 THROUGH 3-33, REFER TO FIGURE 3C.

3-28. Assume that the switch has been moved from position 4 to the position shown. Which of the following conditions now exists?

1. An electrostatic field exists between the capacitor's plates
2. No potential difference exists across the capacitor
3. Current flow is at its maximum
4. Energy is being stored in the capacitor's electrostatic field

3-29. To charge the capacitor, the switch must be in what position?

1. 1
2. 2
3. 3
4. 4

3-30. Which of the following are paths for current flow when the capacitor is charging?

1. Plate A, Plate B, Batt (+) and -Batt (−), Plate A
2. Batt (+) and -Batt (−), Plate A Batt (−) and Batt (+), Plate B
3. Batt (−), Plate A and Plate B, Batt (+), Batt (−)
4. Batt (+), Plate B and Plate A, Batt (−), Batt (+)

3-31. When the switch is placed in position 4, after being in position 2, which of the following conditions exits within the circuit?

1. E_C is increasing
2. I_C is increasing
3. Electrical energy is stored in the capacitor
4. Stored energy is returned to the circuit

3-32. With S1 in position 4, which of the following is the path for current flow?

1. Plate B, Plate A, S1, Plate B
2. Plate A, Plate B, S1, Plate A
3. Plate A, S1, Plate B
4. Plate B, S1, Plate A

3-33. If the illustrated capacitor has a value of 50 pF and a potential difference of 300 volts exists across its plates, what is the total number of coulombs it contains?

1. 0.015
2. 0.15
3. 1.50
4. 15.0

3-34. The greatest rate of change in current occurs between what two times?

1. $T_1 - T_2$
2. $T_2 - T_3$
3. $T_4 - T_5$
4. $T_0 - T_1$

3-35. At what instant does the maximum voltage appear across the resistor?

1. T_1
2. T_2
3. T_5
4. T_0

3-36. When the charge on the capacitor is equal to 100 volts, what is the voltage drop across the resistor?

1. 100 volts
2. 63 volts
3. 27 volts
4. 0 volts

3-37. After the capacitor has reached full charge, S1 is placed in position 2. The greatest rate of change in current is between what two times?

1. $T_1 - T_2$
2. $T_2 - T_3$
3. $T_4 - T_5$
4. $T_0 - T_1$

3-38. The capacitor will be completely discharged at what minimum time interval?

1. T_1
2. T_5
3. T_3
4. T_4

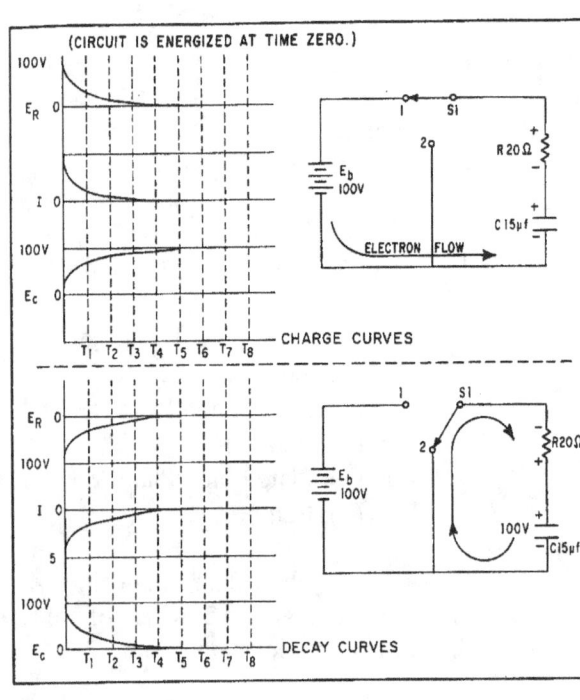

Figure 3D.—RC charge and discharge circuits.

IN ANSWERING QUESTIONS 3-34 THROUGH 3-41, REFER TO FIGURE 3D.

3-39. What is the RC time constant for the circuit?

1. 300 sec
2. 35 sec
3. 300 μsec
4. 35 μsec

3-40. What total time will it take the capacitor to charge to 98 volts? (You may use figure 3-11 of your text, or figure 3 located on this page.)

1. 140 μsec
2. 1200 μsec
3. 140 sec
4. 1200 sec

3-41. After the capacitor has reached full charge, S1 is moved to position 2. What total number of RC time constants will it take for the capacitor to discharge to 5 volts?

1. One
2. Two
3. Three
4. Four

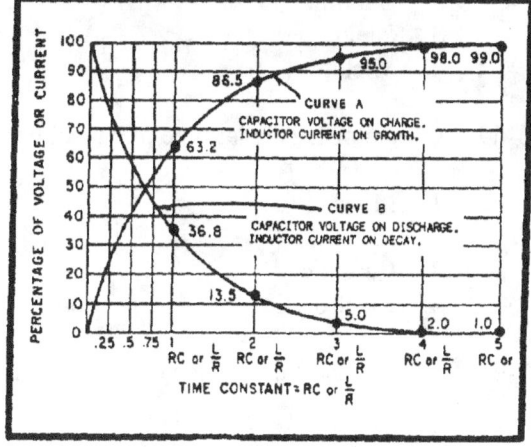

Figure 3E.—Universal time constant chart for RC and RL circuit.

IN ANSWERING QUESTION 3-42, REFER TO FIGURE 3E ABOVE.

3-42. An RC circuit is designed in which a capacitor must charge to 55 percent (.55) of the maximum charging voltage in 200 microseconds. The resistor has a value of 30,000 ohms. What value of capacitance is needed?

1. 0.0089 pF
2. 89.0 pF
3. 0.0089 μF
4. 89.0 μF

MATCH THE CAPACITOR CONFIGURATION IN COLUMN B WITH THE CHARACTERISTICS IN COLUMN A.

A. CHARACTERISTICS B. CONFIGURATION

3-43. Increases total capacitance 1. Capacitors in parallel
3-44. Effectively moves plates further apart 2. Capacitors in series
3-45. Increases plate area
3-46. Total capacitance is found by adding all capacitances
3-47. Decreases total capacitance
3-48. Similar to resistors in parallel

3-49. A circuit contains four parallel-connected capacitors of 33 μF each. What is the total capacitance of the circuit?

1. 8.3 μF
2. 33.0 μF
3. 183.0 μF
4. 132.0 μF

3-50. A circuit contains two series-connected capacitors of 15 μF, and 1500 pF. What is the total capacitance of the circuit?

1. 0.0015 pF
2. 150.0 pF
3. 0.0015 μF
4. 0.1500 μF

3-51. A circuit contains two 10 µF capacitors wired together in a parallel configuration. The two parallel-wired capacitors are wired in series with a 20 µF capacitor and a 20 K ohm resistor. Which of the following expresses the RC time constant for this circuit?

1. .2 sec
2. 2 sec
3. 20,000 sec
4. Both 2 and 3 above

3-52. How are fixed capacitors classified?

1. By their plate size
2. By their dielectric materials
3. By the thickness of their dielectric materials
4. By the thickness of their conductors

3-53. Which of the following types of capacitors are referred to as self-healing?

1. Ceramic
2. Paper
3. Oil
4. Mica

MATCH THE CAPACITOR TYPE IN COLUMN B WITH THE CHARACTERISTIC IN COLUMN A.

	A. CHARACTERISTIC		B. TYPE
3-54.	Has an oxide film dielectric	1.	Electrolytic
3-55.	Can be adjusted by a screw setting	2.	Trimmer
3-56.	A polarized capacitor	3.	Mica
3-57.	Has a waxed paper dielectric	4.	Paper
3-58.	An adjustable capacitor with a mica dielectric		

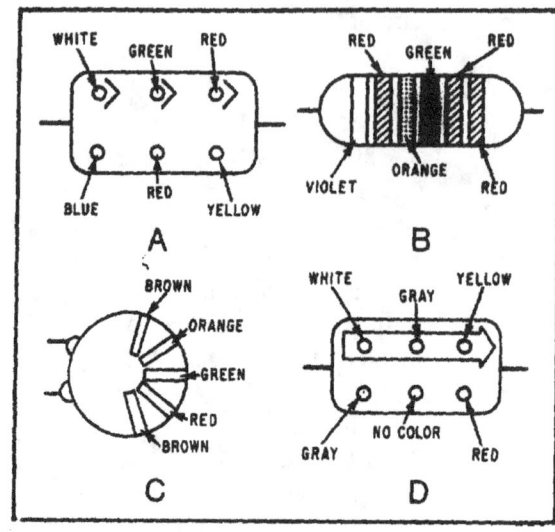

Figure 3F.—Types of capacitors.

IN ANSWERING QUESTIONS 3-59 THROUGH 3-63, REFER TO FIGURE 3F AND TO THE ASSOCIATED PAGES IN YOUR TEXTBOOK.

3-59. Capacitor A is what type of capacitor?

1. Electrolytic
2. Ceramic
3. Paper
4. Mica

3-60. Capacitor B is what type of capacitor?

1. Mica
2. Paper
3. Ceramic
4. Electrolytic

3-61. What is the capacitance of capacitor B?

1. 2,200,000 pF
2. 2,200,000 µF
3. 72,000 pF
4. 72,000 µF

3-62. What is the (a) temperature coefficient and (b) multiplier of capacitor C?

1. (a) -30   (b) 100
2. (a) -30   (b) 1000
3. (a) -330   (b) 100
4. (a) -330   (b) 1000

3-63. What is the (a) capacitance, and (b) voltage rating of capacitor D?

1. (a)   4800 µF   (b) 200 volts
2. (a)   4800 pF   (b) 200 volts
3. (a)   98,000 µF   (b) 800 volts
4. (a) 980,000 pF   (b) 800 volts

ASSIGNMENT 4

Textbook assignment: Chapter 4, "Inductive and Capacitive Reactance," pages 4-1 through 4-40. Chapter 5, "Transformers," pages 5-1 through 5-31.

4-1. Inductance has what effect, if any, on a change in (a) current, and (b) voltage?

1. (a) No effect (b) aids it
2. (a) Aids it (b) no effect
3. (a) Opposes it (b) no effect
4. (a) No effect (b) opposes it

4-2. Voltage leads current in which of the following types of circuits?

1. Resistive
2. Capacitive
3. Both 1 and 2 above
4. Inductive

4-3. Opposition to the flow of current by a coil in an ac circuit is represented by what symbol?

1. R
2. X_L
3. L
4. H

4-4. What is the opposition offered by a coil to (a) the flow of alternating current and (b) a change in current?

1. (a) Resistance (b) Inductance
2. (a) Reactance (b) Counterreactance
3. (a) Reactance (b) Inductance
4. (a) Resistance (b) Reactance

4-5. The formula $2\pi fL$ is used to determine what electrical quantity?

1. Resistance
2. Inductance
3. Counterreaction
4. Inductive reactance

4-6. An inductive circuit contains a 200-µH coil and the ac voltage applied is at a frequency of 120 Hz. What is the value of reactance for the circuit?

1. 0.15 Ω
2. 1.50 Ω
3. 7.50 Ω
4. 75.0 Ω

4-7. If the frequency applied to a circuit with a 200-µH coil is increased from 120 Hz to 50 kHz, what will be the value of reactance for the circuit?

1. 1.0.75 Ω
2. 2.7.5 Ω
3. 3.62.8 Ω
4. 628.0 Ω

4-8. A capacitor will (a) conduct what type of current, and (b) block what type of current?

1. (a) Dc (b) All ac
2. (a) All ac (b) Dc
3. (a) Dc (b) Ac above 60 Hz
4. (a) Ac above 60 Hz (b) Dc

THIS SPACE LEFT BLANK INTENTIONALLY.

24

IN ANSWERING QUESTIONS 4-9 THROUGH 4-13, SELECT FROM COLUMN B THE PROPERTY THAT CAUSES THE ELECTRICAL EFFECT IN COLUMN A.

	Column A		Column B
4-9.	Opposition to ac but not dc	1.	Inductive reactance
4-10.	Causes a phase shift between voltage and current	2.	Capacitive reactance
4-11.	Increases with an increase in frequency	3.	Both 1 and 2 above
4-12.	Causes current to lead voltage by 90°	4.	Resistance
4-13.	Decreases with an increase in frequency.		

4-14. An electrical circuit contains a 25-µF capacitor and operates from a 60-Hz ac source. What is the value of capacitive reactance of the circuit?

1. 0.00106 Ω
2. 0.0106 Ω
3. 10.6 Ω
4. 106.2 Ω

THIS SPACE LEFT BLANK INTENTIONALLY.

IN ANSWERING QUESTIONS 4-15 AND 4-16 USE THE FOLLOWING INFORMATION: A SERIES CIRCUIT HAS AN INDUCTIVE REACTANCE OF 56Ω, A CAPACITIVE REACTANCE OF 25Ω, AND OPERATES AT A FREQUENCY OF 400 HZ.

4-15. What formula should you use to determine the total reactance for the circuit?

1. $X = 2\pi fL$
2. $X = \dfrac{1}{2\pi fC}$
3. $X = X_L - X_C$
4. $X = X_C - X_L$

4-16. What is the total value of reactance for the circuit?

1. 31 Ω
2. 81 Ω
3. 1,400 Ω
4. 14,067 Ω

4-17. What term is used to express the total opposition to ac in an electrical circuit?

1. Reactance
2. Impedance
3. Resistance
4. Conductance

4-18. A series ac circuit has the following values: $X_L = 5\Omega$, $X_C = 6\Omega$, and $R = 7\Omega$. What is the value of Z?

1. Ω
2. 3.03 Ω
3. 7.07 Ω
4. 14.14 Ω

4-19. A series circuit contains an inductor having 12 ohms of resistance and 30 ohms of inductive reactance in series with a capacitor having 21 ohms of capacitive reactance. The applied voltage is 100 volts. What is the value of current for the circuit?

1. 6.6 amps
2. 8.4 amps
3. 15.0 amps
4. 25.6 amps

4-20. A series circuit contains an inductor having 12 ohms of resistance and 64 ohms of inductive reactance in series with a capacitor having 69 ohms of capacitive reactance. If the current through the circuit is 6.5 amperes, what is the value of the voltage applied to the circuit?

1. 26.5 volts
2. 55.5 volts
3. 75.5 volts
4. 84.5 volts

4-21. True power in a circuit is dissipated in what circuit element?

1. Resistance
2. Reactance
3. Capacitance
4. Inductance

4-22. In a purely reactive circuit, what happens to power?

1. It is dissipated across the reactive loads
2. It is cancelled by the reactive elements
3. It is stored in the reactive elements
4. It is returned to the source

4-23. True power is measured in what unit?

1. Watt
2. Volt-ampere
3. Var
4. P_t-watt

4-24. An ac series circuit has the following characteristics: R = 8 ohms, X_C = 100 ohms, X_L = 70 ohms, and E = 220 V. What is the value of true power for the circuit?

1. 46 W
2. 57 W
3. 268 W
4. 402 W

4-25. What is the unit of measurement for reactive power?

1. Watt
2. Var
3. Volt-ampere
4. Volt-ohm

4-26. An ac series circuit has the following values: I = 7.5 amps, X_L = 80Ω, and X_C = 35Ω. What is the value of reactive power for the circuit?

1. 2531 var
2. 1567 var
3. 1283 var
4. 861 var

4-27. Apparent power in an ac circuit is a combination of which of the following factors?

1. Applied power and true power
2. Reactive power and true power
3. Applied power and the power returned to the source
4. Reactive power and the power returned to the source

4-28. What is the unit of measurement for apparent power?

1. Watt
2. Var
3. Volt-ampere
4. Volt-ohm

4-29. An ac circuit dissipates 800 watts across its resistance and returns 600 var to the source. What is the value of the apparent power of the circuit?

1. 200 VA
2. 500 VA
3. 1000 VA
4. 1400 VA

4-30. The portion of apparent power dissipated in an ac circuit can be calculated by which of the following formulas?

1. $PF = (I_R)^2 R$
2. $PF = (I_Z)^2$
3. $PF = \dfrac{(I_Z)^2 Z}{(I_Z)^2 R}$
4. $PF = \dfrac{(I_R)^2 R}{(I_Z)^2 Z}$

4-31. A series ac circuit has a X_C of 110 ohms, an X_L of 30 ohms, and a circuit resistance of 22 ohms. What is the power factor of this circuit?

1. .91
2. .27
3. .20
4. .13

An RLC series a.c. circuit has the following values:

E = 65 volts
f = 120 Hz
R = 12 ohms
L = 30 mH
C = 450 μF

Figure 4A.—Circuit characteristics.

IN ANSWERING QUESTIONS 4-32 THROUGH 4-36, REFER TO FIGURE 4A.

4-32. What is the value of X?

1. 19.7 Ω
2. 27.8 Ω
3. 31.6 Ω
4. 42.3 Ω

4-33. What is the value of Z?

1. 23 Ω
2. 28 Ω
3. 33 Ω
4. 38 Ω

4-34. What is the value of I_T for the circuit?

1. 1.8 A
2. 2.8 A
3. 3.4 A
4. 4.4 A

4-35. What is the value of true power?

1. 67 W
2. 83 W
3. 94 W
4. 125 W

4-36. What is the power factor?

1. .46
2. .52
3. .73
4. .88

4-37. When impedance is calculated for a parallel ac circuit, an intermediate value must first be calculated. The intermediate value must then be divided into the source voltage to derive impedance. What is this intermediate value?

1. Reactance
2. Resistance
3. Power factor
4. Total current

4-38. Which of the following defines transformer action?

1. The transfer of energy from one circuit to another through electromagnetic action
2. The transfer of energy from one circuit to another through electrostatic action
3. The development of counter electromotive force where a magnetic field cuts a coil
4. The development of a voltage across a coil as it cuts through a magnetic field

4-39. Which of the following is NOT a necessary element in a basic transformer?

1. A core
2. A primary winding
3. A secondary winding
4. A magnetic shield

4-40. What three materials are most commonly used for transformer cores?

1. Copper, soft iron, and air
2. Copper, soft iron, and steel
3. Air, copper, and steel
4. Air, soft iron, and steel

4-41. The two types of transformer cores most commonly used are the shell-core and the

1. I-core
2. E-core
3. hollow-core
4. laminated-core

4-42. What is the major difference between the primary and secondary windings of a transformer?

1. The primary has more turns than the secondary
2. The secondary has more insulation than the primary
3. The primary is connected to the source; the secondary is connected to the load
4. The primary is connected to the load; the secondary is connected to the source

4-43. What is the principal difference between a high-voltage transformer and a low-voltage transformer?

1. A high-voltage transformer has more turns of wire than a low-voltage transformer
2. A high-voltage transformer uses a hollow-core, while a low-voltage transformer uses a shell-type core
3. A high-voltage transformer uses a shell-type core, while a low-voltage transformer uses a hollow-core
4. A high-voltage transformer has more insulation between the layers of windings than does a low-voltage transformer

THIS SPACE LEFT BLANK INTENTIONALLY.

IN ANSWERING QUESTIONS 4-44 THROUGH 4-46, SELECT FROM COLUMN B THE TRANSFORMER CHARACTERISTICS THAT ARE IDENTIFIED IN THE TRANSFORMER SCHEMATICS IN COLUMN A.

A. SCHEMATIC   B. TRANSFORMER TYPE

4-44.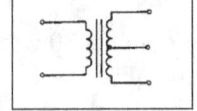
1. Air-core
2. Iron-core
3. Center-tapped
4. Iron-core with center tap

4-45.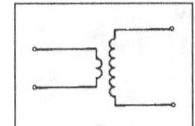

4-46.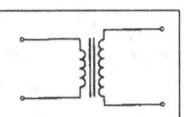

4-47. When the secondary of a transformer is NOT connected to a circuit, the transformer is said to be operating under which of the following conditions?

1. Uncoupled
2. No-load
3. Loaded
4. Open

4-48. What term applies to the current in the primary of a transformer that creates the magnetic field?

1. Exciting current
2. Primary current
3. Magnetizing current
4. Counter current

4-49. In the primary of a transformer, what opposes the current from the source?

1. The impedance
2. The forward emf
3. The self-induced emf
4. The exciting current

4-50. What is the source of the magnetic flux that develops secondary voltage in a transformer?

1. Primary emf
2. Secondary counter emf
3. Primary exciting current
4. Secondary exciting current

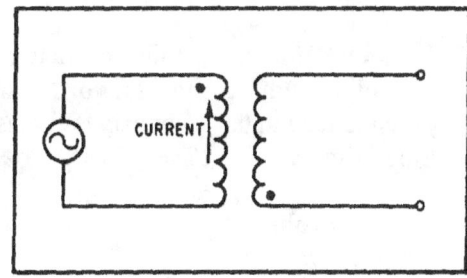

Figure 4B.—Transformer polarity.

IN ANSWERING QUESTION 4-51, REFER TO FIGURE 4B.

4-51. The illustrated transformer is (a) what type, and (b) in what direction is the current flowing in the secondary?

1. (a) Like-wound   (b) ↑
2. (a) Unlike-wound (b) ↑
3. (a) Like-wound   (b) ↓
4. (a) Unlike-wound (b) ↓

4-52. Which of the following terms applies to the flux from the primary that does NOT cut the secondary

1. Lost flux
2. Leakage flux
3. Uncoupled flux
4. Coefficient flux

4-53. What is the main cause for the coefficient of coupling of a transformer being less than unity?

1. Counter emf
2. Induced emf
3. Uncoupled flux
4. Leakage flux

4-54. A transformer has a source voltage of 50 volts ac, with a turns ratio of 1:6. The coefficient of coupling is 1.0. What is the voltage of the secondary winding?

1. 150
2. 300
3. 500
4. 600

4-55. A transformer has a unity coefficient of coupling with a 5:1 turns ratio; 20 volts are induced in the secondary. What is the primary voltage?

1. 100 volts
2. 50 volts
3. 10 volts
4. 4 volts

4-56. A transformer has a unity coefficient of coupling. Thirty-five volts applied to its primary induces 105 volts in its secondary. The secondary is composed of 99 turns. What is the number of turns in the primary?

1. 11 turns
2. 22 turns
3. 33 turns
4. 44 turns

4-57. A transformer secondary has 20 amperes of current flowing at 60 volts potential. The applied voltage is 10 volts. What is (a) the turns ratio of the transformer and (b) what total current is flowing in the primary?

1. (a) 6:1, (b) 3.3 amperes
2. (a) 1:6, (b) 120 amperes
3. (a) 1:2, (b) 10 amperes
4. (a) 2:1, (b) 120 amperes

4-58. A 2:1 transformer delivers 30 watts to the load and 3 watts of power are lost to internal losses. What total power is drawn from the source?

1. 63 watts
2. 57 watts
3. 33 watts
4. 27 watts

4-59. What is the efficiency of the transformer described in question 4-58?

1. 33 %
2. 46 %
3. 53 %
4. 91 %

THIS SPACE LEFT BLANK INTENTIONALLY.

IN ANSWERING QUESTIONS 4-60 THROUGH 4-62, SELECT FROM COLUMN B THE TERM THAT DESCRIBES THE TYPE OF POWER LOSS IN COLUMN A.

	A. LOSS TYPE		B. TERMS
4-60.	Power lost in realigning domains	1.	Copper loss
4-61.	Power dissipated by the resistance of the windings	2.	Eddy-current loss
4-62.	Power loss caused by random core currents	3.	Hysteresis loss
		4.	Leakage Loss

4-63. A transformer designed for a low frequency will NOT be damaged when used at higher frequencies. What change within the transformer, limits transformer current to a safe value at higher frequencies?

1. Increased hysteresis loss
2. Increased inductive reactance
3. Increased leakage flux
4. Increased eddy-current loss

THIS SPACE LEFT BLANK INTENTIONALLY.

IN ANSWERING QUESTIONS 4-64 THROUGH 4-67, SELECT THE TRANSFORMER TYPE FROM COLUMN B THAT PERFORMS THE TASK OR HAS THE CHARACTERISTICS DESCRIBED IN COLUMN A.

	A. TASK		B. TRANSFORMER TYPE
4-64.	Used above 20 kHz	1.	Power
4-65.	The secondary is a tapped primary	2.	Autotransformer
4-66.	Used to deliver voltage from a source to a load	3.	Audio-Frequency
4-67.	Can be used to match impedance in a sound system	4.	Radio-Frequency

4-68. What wire colors conventionally identify the secondary center tap of a power transformer?

1. Black and yellow
2. Red and white
3. Black and red
4. Red and yellow

4-69. Before starting to work on any electrical equipment, you should first determine that the equipment is in which of the following conditions?

1. Connected
2. Deenergized
3. Energized
4. Operational

4-70. A person is working on electrical equipment. The power is secured and tagged. The technician receives a shock on the hand. What safety precaution was overlooked?

1. The technician was not standing on approved rubber matting
2. The technician had not discharged the equipment's capacitors
3. The technician was working on energized equipment
4. The technician had two hands in the equipment

4-71. When working on electrical equipment, why should you use only one hand?

1. The free hand can be used to turn off the power in case of shock
2. The free hand can be used to pull the other hand free in case of muscle contraction from shock
3. The free hand will ensure that you are properly grounded
4. The free hand will minimize the possibility of creating a low resistance path to ground through your body

www.ingramcontent.com/pod-product-compliance
Lightning Source LLC
Chambersburg PA
CBHW081143180526
45170CB00006B/1908